K.B. Newman • S.T. Buckland • B.J.T. Morgan
R. King • D.L. Borchers • D.J. Cole • P. Besbeas
O. Gimenez • L. Thomas

Modelling Population Dynamics

Model Formulation, Fitting and Assessment using State-Space Methods

Springer

K.B. Newman
Stockton Fish and Wildlife Office
Pacific Southwest Region
U.S. Fish and Wildlife Service
Lodi, CA, USA

B.J.T. Morgan
D.J. Cole
School of Mathematics, Statistics and Actuarial Science
University of Kent
Canterbury, Kent, UK

O. Gimenez
Centre d'Écologie Fonctionnelle et Évolutive, UMR 5175
Campus du CNRS
Montpellier cedex 5
France

S.T. Buckland
R. King
D.L. Borchers
L. Thomas
CREEM
The Observatory
Buchanan Gdns
St. Andrews, UK

P. Besbeas
Department of Statistics
Athens University of Economics and Business
Athens, Greece
And
School of Mathematics, Statistics and Actuarial Science
University of Kent
Canterbury, Kent, UK

ISBN 978-1-4939-0976-6 ISBN 978-1-4939-0977-3 (eBook)
DOI 10.1007/978-1-4939-0977-3
Springer New York Heidelberg Dordrecht London

Library of Congress Control Number: 2014939376

Printed on acid-free paper

Springer is part of Springer Science+Business Media (www.springer.com)

Preface

In Borchers et al. (2002), a unifying framework was given for the plethora of methods for estimating the abundance of closed populations of animals. In Chap. 13 of that book, the bare bones of an approach to provide similar unification for open populations were provided. Here, we seek to put flesh on those bones. Our aims are to lay down an approach for defining and fitting a wide range of population dynamics models for those new to the field and to provide some coherence to the process of modelling population dynamics for more experienced practitioners.

Lodi, CA, USA	K.B. Newman
St. Andrews, UK	S.T. Buckland
Canterbury, UK	B.J.T. Morgan
St. Andrews, UK	R. King
St. Andrews, UK	D.L. Borchers
Canterbury, UK	D.J. Cole
Athens, Greece and Canterbury, UK	P. Besbeas
Montpellier, France	O. Gimenez
St. Andrews, UK	L. Thomas

Contents

Authors' (Brief) Biographies

Ken Newman is a mathematical statistician for the United States Fish and Wildlife Service. His primary research interest is in the use of state-space models to model the population dynamics of fish and wildlife to answer scientific questions and to serve as management decision support tools.

Stephen Buckland is professor of statistics at the University of St Andrews and is also co-director of the UK National Centre for Statistical Ecology. His interest in modelling population dynamics dates from a project to develop a decision support system for managers of red deer populations in Scotland in the 1990s.

Byron Morgan is honorary professorial research fellow in the University of Kent and co-director of the National Centre for Statistical Ecology. He is interested in integrated population modelling, which accounts for data collected on different aspects of the demography of wild animals. A convenient component of this work is the use of state-space models for describing ecological time series.

Ruth King is a reader in statistics at the University of St Andrews. Her research interests include the development of population dynamics models and model fitting tools in both the classical and Bayesian frameworks. This particularly includes the application of hidden Markov (or state-space) models.

David Borchers is a reader in statistics at the University of St Andrews. His research involves developing general statistical models for estimating population density and distribution, integrating hidden state or latent variable models with various kinds of observation model.

Diana Cole is a senior lecturer in statistics at the University of Kent. Her primary research is on parameter redundancy or identifiability of models used in statistical ecology.

Panagiotis Besbeas is a lecturer in the Athens University of Economics and Business and also a part-time post-doctoral research associate within the National Centre for Statistical Ecology group in the University of Kent, Canterbury. His research includes integrated population modelling, recently including the importance of replication for error estimation, and new methods for measuring goodness of fit as well as for conducting model selection.

Olivier Gimenez is a senior scientist in statistical ecology at the Centre National de la Recherche Scientifique (CNRS) in France. His main research interest is animal demography using hidden structure models with contributions to the coexistence of humans and animals.

Len Thomas is a reader in statistics at the University of St Andrews and is director of the Centre for Research into Ecological and Environmental Modelling, an interdisciplinary research centre at the university. He has two main research interests: (1) use of computer-intensive methods, particularly particle filters, to fit and compare state-space models of wildlife population dynamics; (2) development of methods and software for estimating animal population size and density.

We thank the following institutions for their support in the preparation of this book: US Fish and Wildlife Service (Ken B. Newman), University of St. Andrews (Stephen T. Buckland, Ruth King, David L. Borchers and Len Thomas), University of Kent (Byron J.T. Morgan, Diana J. Cole and Panagiotis Besbeas), the Athens University of Economics and Business (Panagiotis Besbeas) and CNRS (Olivier).

Chapter 1
Introduction

This book was conceived as a companion volume to *Estimating Animal Abundance: Closed Populations* by Borchers et al. (2002), who gave a unifying framework for estimating the abundance of closed populations. We seek to do the same for estimating the abundance of open populations: populations subject to births, deaths and movement in and out of the population. We focus primarily on populations of large vertebrates, for which we will typically model dynamics within the framework of an annual cycle, and for which stochastic variability in the demographic processes is usually modest. We consider discrete time models in which animals can be assigned to discrete states such as age class, gender, population (within a metapopulation), or species (for multi-species models).

We go beyond estimation of abundance, however, as we are interested in understanding reasons for variation in abundance over time. Knowledge of abundances alone will seldom be sufficient for the ecologist or wildlife manager. For example, suppose one knew there were exactly 5,218, 6,319, and 7,438 individuals for three consecutive years. A natural question would be "*Why* are the abundances increasing?". More generally, the question is "What is causing, or driving, the population dynamics?". The answer will depend in some way on the underlying processes of birth or recruitment, survival and movement. These processes are typically of greater interest to ecologists and managers than abundance itself.

To quantify these processes requires the formulation and fitting of population dynamics models. The resulting fitted models will potentially yield both estimates of abundance and estimates of parameters characterizing the underlying processes, such as survival probabilities.

1.1 Background to the Book

Population dynamics models serve a wide variety of uses, including explaining previous population fluctuations as well as projecting future population abundances under hypothetical scenarios. We might for example wish to reintroduce an animal

K.B. Newman et al., *Modelling Population Dynamics: Model Formulation, Fitting and Assessment using State-Space Methods*, Methods in Statistical Ecology,

DOI 10.1007/978-1-4939-0977-3_1,

that has become extinct locally. We then need to be confident that the reintroduced population is viable. Models are needed to make this assessment. Or perhaps we need to manage an exploited fish stock, and assess whether the level of take is sustainable. Another example is when we wish to manage a species of conservation concern, and perhaps reverse a declining trend. We would like models of the species and its environment to allow us to explore "what-if" scenarios on the computer before we experiment with the real population. If we can build a model that relates demographic processes to habitat, we can then explore options for modifying the habitat to the species' benefit. Perhaps we wish to assess the response of a species or community to climate change. To do this, we must model the population processes, and how these depend on climate. Yet another example is when a population is too large for the environment to support, so that habitat degrades, to the detriment of biodiversity in the region. We may then need models to indicate levels of culling needed to sustain the environment and its diversity.

Population dynamics models play a central role in adaptive resource management (Williams et al. 2007) where managers are experimenting with different management actions to produce desired population responses. Models are used to predict the consequences of management actions such as habitat modifications or predator removal. After-the-fact comparisons of model predictions to observed outcomes are also made. The strength of evidence for alternative models, i.e. competing explanations for the effects of actions, can be assessed to determine which explanations are more consistent with the data.

In all of the above examples, we need to address risk. This means that we need to quantify uncertainty in our model predictions. It is not enough therefore to develop mathematical models of animal dynamics. We also need to allow for the main sources of uncertainty in these models: demographic and environmental stochasticity; observation or sampling errors in our data; and model or structural uncertainty (Williams et al. 2001). Without these elements, a mathematical model of population dynamics is merely a tool of theoretical ecologists, with little to offer wildlife managers. Similarly, purely empirical models for trend, fitted to time series of data, have little predictive power, because they ignore the population processes that drive change, so cannot predict change with any degree of confidence. In this book, we develop an approach that allows models for population dynamics to be embedded within a full inferential framework, giving the predictive and explanatory power of mathematical process models, coupled with statistical tools for quantifying uncertainty.

The main steps involved are as follows. First, we formulate a model of the population dynamics. We provide a building block strategy for this, allowing quite complex models to be assembled from simple components. For convenience of model formulation, we use matrices to define these building blocks, which allows us also to relate our approach to the well-developed field of matrix population models (Caswell 2001). However, our fitting algorithms are not restricted to such models, and we can readily extend our models in various ways. We develop the tools needed for model formulation, and show how they provide a more general framework than is possible with matrix models.

Second, we formulate the population dynamics model as a hidden process model. If the model is first-order Markov, it is termed a *state-space model*. The states might reflect gender, age, geographic region, or other characteristics that define different sub-populations that might be subject to different dynamics. The state equation defines how the population updates stochastically through time. These processes are generally hidden because we cannot observe the entire population—we typically have counts or estimates of the population or of some component of the population at certain points in time. Hidden process models give great flexibility in defining both the systematic and the random components of the population dynamics model.

Third, we define how our observations relate to the states, through an observation equation. The random component of this equation relates to observation error, while the systematic component defines a "mapping" between the observations and the states.

As an aside, we note that partitioning the population dynamics model into two components, namely the underlying hidden state process and the observation model, has practical advantages. One advantage is that theories about the dynamics can be kept separate from the sampling and estimation issues; thus, for example, the subject matter specialist can focus attention on the underlying dynamic processes of survival and birth, postulating alternative theories largely independent of the data collection procedures. Sample size determination can also be partitioned into problems to determine the number of years to sample and problems to determine the sampling intensity within a given year, and trade-offs between different combinations of sample sizes can be more clearly determined.

Fourth, we address how the hidden process model may be fitted to time series of observations. We consider three primary approaches. The Kalman filter allows us to fit a model quickly, at the expense of assuming that the errors in both the state equation and the observation equation are normally distributed, and that the expectation of the state vector at one time point is a linear function of the state vector at the previous time point. We also give two computer-intensive methods which can be implemented within a Bayesian framework to give much greater flexibility. The first of these is Markov chain Monte Carlo (MCMC), and the second is sequential Monte Carlo methods.

Finally, we discuss model selection, goodness-of-fit, model averaging, parameter redundancy, and other issues related to model assessment and uncertainty.

In subsequent chapters, we consider different methods of assessing wild animal populations, and how these can be combined with hidden process modelling to allow the dynamics of the populations to be modelled. We first address the case where closed-population assessment methods are used at each of several time points, and a population dynamics model fitted to the time series of abundance estimates. This may be viewed as an extension of the robust design approach of Pollock (1982). Next, we consider open population assessment methods, and how these may be integrated with hidden process models. Finally, we look at the case when there are multiple data sources. These methods are illustrated through examples, and `R` or `WinBUGS` code is provided for most examples, allowing users to modify the code for their own purposes.

1.2 Book Website

Computer code for our examples is available at the book website, www.creem.st-and.ac.uk/modpopdyn. The code can be used as a learning aid to re-run the analyses of this book, and can also be used as a template for the user to generate code to analyse his or her own data.

1.3 Related Books

As noted above, this book is a companion to Borchers et al. (2002). They give a unifying framework for methodologies for estimating the abundance of closed populations, and we seek to do the same for open populations. Population dynamics play a crucial role in open population modelling, yet methods for assessing the abundance of open populations have for the most part had no embedded model of these dynamics. Such embedded models are central to the methods of this book.

This book also complements the book on Bayesian analysis for population ecology by King et al. (2009). They focus on the underlying ideas associated with Bayesian analyses of ecological data and the corresponding methods to fit Bayesian models. There is an emphasis on mark-recapture-recovery data for open populations. Case studies are used to illustrate the methods, including state-space models, the use of covariate information (including dealing with missing data), multi-state data, integrated data analyses, random effects models and closed capture-recapture models. Computer codes are provided for examples, using both `WinBUGS` and `R`.

Williams et al. (2001) offer near encyclopedic coverage of many of the better known population dynamics models for animal populations as well as the wide variety of statistical procedures used to estimate abundance (for closed and open populations) and parameters of population dynamic processes. Integration of population process models with observation models is by and large not addressed, however.

Caswell (2001) uses matrices to model population dynamics, e.g., Leslie matrices to characterise survival and reproduction. Matrices, strictly speaking, imply linear models, although Caswell formulates non-linear extensions to the basic matrix model. We find matrix models to be useful for formulating state-space models and present many examples in Chap. 2.

Royle and Dorazio (2008) cover hierarchical models for ecological systems. In common with this book, they use both frequentist and Bayesian methods, and show how to combine explicit models of ecological system structure or dynamics with models of how ecological systems are observed. They apply the principles of hierarchical modelling to problems in population, metapopulation, community and metacommunity systems, whereas we focus on modelling population dynamics, and estimating population abundance.

Two classic fisheries science and management books, Hilborn and Walters (1992) and Quinn and Deriso (1999), describe a wide variety of methods and models for characterising fish population dynamics. These models and methods are often used for non-fish species. Both books also discuss the distinction between process variation and observation error, but at the time the books were written, the handling of both types of error simultaneously was not common practice.

Written for researchers and graduate students in statistics, ecology, demography, and the social sciences, McCrea and Morgan (2014) provide an up-to-date overview of capture-recapture methods. It covers model development and diagnostics. The authors use illustrative data sets drawn from a range of different areas and indicate available software for classical and Bayesian inference.

We use Bayesian methods for fitting many of the models of this book. Books that provide an introduction to Bayesian methods for ecologists are McCarthy (2007) and Link and Barker (2009). Kéry and Schaub (2012) focus on using `WinBUGS` to model population dynamics. They show how to fit Bayesian state-space models using mark-recapture or mark-recovery data, and they also provide example code for integrated population modelling, and for modelling counts.

1.4 Roadmap to the Book

After reading this chapter, we recommend that Chap. 2 be read next, which provides a path for those familiar with classic Leslie and Lefkovitch matrix models to a building-block approach to (a) constructing situation-specific matrix models and (b) formulating the skeleton of state-space models. Chapters 3–5 are the methodological core of the book giving general definitions of SSMs, methods for fitting SSMs, and then suggestions for formulation, or re-formulation, and assessment of SSMs. Chapters 6–9 are largely stand-alone chapters and can be read in any order, although there is some cross-reference between these chapters. Chapter 10 summarizes the book and can be read at any time.

Chapter 2
Matrix Models as Building Blocks for Population Dynamics

In this chapter, we develop a "building block" approach to defining population dynamics models, in which each building block corresponds to one biological process, and is represented by one matrix (Lebreton 1973; Lebreton and Isenmann 1976; Buckland et al. 2004, 2007). Matrix models are usually defined within a deterministic framework, but we will need stochastic models. Thus we will think of the matrix as a mathematical tool for telling us how many animals of each type we expect to have in our population once the process (e.g. survival, birth, movement) has occurred, given the numbers present beforehand. We separately specify the probability distribution associated with the process, which will determine the actual numbers of each type. We refer to the types of animal as *states*.

We will introduce each process using a simple example to aid understanding, then define it in more general terms. We will assume that the natural time unit is one year, although a shorter unit will often be appropriate. We will also assume that births are restricted to a short breeding season. The year is taken to run from one breeding season to the next. If the breeding season is not short, then the birth process operates alongside a death process, which would be better represented by continuous-time models. We can approximate such models in our framework by modelling births and deaths on say a daily basis through the breeding season.

Within a year, multiple biological processes typically occur, e.g. survival, birth and movement. At times these individual processes will be labelled sub-processes: the annual changes in numbers of animals are a reflection of an annual process, which in turn is the consequence of multiple sub-processes. Further, in the constructions that follow, sub-processes are treated as if they occur sequentially in discrete, non-overlapping time intervals; e.g. mortality takes place, then births occur, then movement, and so on. Temporal overlapping of sub-processes can in fact occur so long as the overlapping sub-processes are affecting different categories or sub-populations; otherwise, the following formulations are approximations.

K.B. Newman et al., *Modelling Population Dynamics: Model Formulation, Fitting and Assessment using State-Space Methods*, Methods in Statistical Ecology,
DOI 10.1007/978-1-4939-0977-3_2,

2.1 Using Matrices to Represent Processes Within a Single Population

Caswell (2001) gives a comprehensive account of matrix population models, and of properties of matrices relevant to modelling population dynamics. In this book, we need to know remarkably little about matrices. The basic property of matrix multiplication is needed to understand the expected effect on states of processes. Suppose A is a 2×2 matrix comprising four values a_{11}, a_{12}, a_{21} and a_{22}, and $\mathbf{b}$ is a column vector of length two (i.e. it is a 2×1 matrix) with elements b_1 and b_2. Then the product $A\mathbf{b}$ is given by

$$\begin{bmatrix} a_{11} & a_{12} \\ a_{21} & a_{22} \end{bmatrix} \begin{bmatrix} b_1 \\ b_2 \end{bmatrix} = \begin{bmatrix} a_{11}b_1 + a_{12}b_2 \\ a_{21}b_1 + a_{22}b_2 \end{bmatrix}.$$

For example

$$\begin{bmatrix} 1 & 2 \\ 3 & 8 \end{bmatrix} \begin{bmatrix} 3 \\ 5 \end{bmatrix} = \begin{bmatrix} 13 \\ 49 \end{bmatrix}.$$

Note that in any product of two matrices, the number of columns of the first matrix must equal the number of rows of the second matrix. The number of rows of the product is equal to the number of rows of the first matrix, and the number of columns of the product is equal to the number of columns of the second matrix. More generally, if $\mathbf{A}$ is a $k \times m$ matrix and $\mathbf{B}$ is an $m \times n$ matrix, then the $k \times n$ matrix $\mathbf{AB}$ is given by:

$$\begin{bmatrix} a_{11} & \dots & a_{1m} \\ \vdots & \ddots & \vdots \\ a_{k1} & \dots & a_{km} \end{bmatrix} \begin{bmatrix} b_{11} & \dots & b_{1n} \\ \vdots & \ddots & \vdots \\ b_{m1} & \dots & b_{mn} \end{bmatrix}$$

$$= \begin{bmatrix} a_{11}b_{11} + \dots + a_{1m}b_{m1} & \dots & a_{11}b_{1n} + \dots + a_{1m}b_{mn} \\ \vdots & \ddots & \vdots \\ a_{k1}b_{11} + \dots + a_{km}b_{m1} & \dots & a_{k1}b_{1n} + \dots + a_{km}b_{mn} \end{bmatrix}.$$

Throughout this book, we use matrices, called *process matrices*, to define the sub-processes operating on our population of interest, and vectors, called *state vectors*, to represent the various categories or *states* of animal in our population. The start of the annual cycle will be defined for convenience, often immediately before or after the breeding season. The numbers of animals in each state at the end of year $t - 1$ will be represented by a vector $\mathbf{n}_{t-1}$, comprising elements $n_{j,t-1}$, for $j = 1, \dots, m$, where m is the number of distinct states of interest. After each sub-process, these numbers (or some of them) will change. We use the following notation to indicate numbers of animals in each of the m states after sub-process k has occurred in year t:

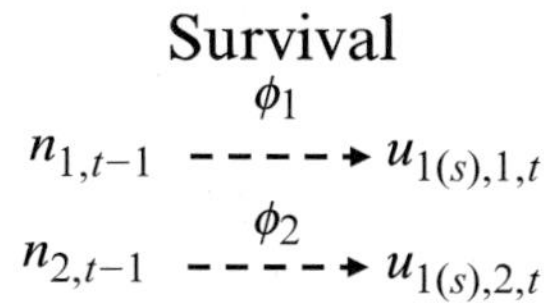

Fig. 2.1 Starting with $n_{1,t-1}$ young animals and $n_{2,t-1}$ adults at the end of year $t-1$, we expect $\mathrm{E}(u_{1(s),1,t}) = n_{1,t-1}\phi_1$ young and $\mathrm{E}(u_{1(s),2,t}) = n_{2,t-1}\phi_2$ adults to survive through year t. The arrows are dashed to indicate that the processes are stochastic, so that in general $\mathrm{E}(u_{1(s),j,t}) \neq u_{1(s),j,t}$ for $j = 1, 2$. The rates associated with the processes are shown above the arrows

$$\mathbf{u}_{k(x),t} = \begin{bmatrix} u_{k(x),1,t} \\ \vdots \\ u_{k(x),m,t} \end{bmatrix}$$

where x is a letter indicating the type of sub-process: s is used to indicate survival, b for birth, a for age incrementation, r for growth, c for sex assignment, g for genotype allocation, and v for movement.

We use the same letters, but capitalized, to label our models. For example a BAS model is one with three sub-processes in a year, starting with survival, then age incrementation, and finally birth. The reason for reversing the chronological order of the sub-processes in these labels will become apparent.

2.1.1 *Survival*

Suppose we wish to model a single population of animals, divided into two states representing age classes, with $n_{1,t-1}$ newly born animals and $n_{2,t-1}$ adults at the end of year $t-1$. Then the expected number of survivors through year t can be expressed as

$$\begin{bmatrix} \mathrm{E}(u_{1(s),1,t}) \\ \mathrm{E}(u_{1(s),2,t}) \end{bmatrix} = \begin{bmatrix} \phi_1 & 0 \\ 0 & \phi_2 \end{bmatrix} \begin{bmatrix} n_{1,t-1} \\ n_{2,t-1} \end{bmatrix}, \tag{2.1}$$

where $u_{1(s),j,t}$ signifies the number of animals in state j after the first sub-process (survival) of year t has occurred, $\mathrm{E}(u_{1(s),j,t})$ denotes the corresponding expectation conditional on $n_{1,t-1}$ and $n_{2,t-1}$, and ϕ_j is the survival probability of animals in state j. The structure of Eq. (2.1) is shown diagrammatically in Fig. 2.1. We write Eq. (2.1) equivalently as $E(\mathbf{u}_{1(s),t}|\mathbf{n}_{t-1}) = \mathbf{S}\mathbf{n}_{t-1}$, where $\mathbf{S} = \begin{bmatrix} \phi_1 & 0 \\ 0 & \phi_2 \end{bmatrix}$ is the survival matrix.

We will also need to specify a probability distribution associated with the process. For survival, an obvious choice is the binomial, so that

$$\begin{pmatrix} u_{1(s),1,t} \sim \text{binomial}(n_{1,t-1}, \phi_1) \\ u_{1(s),2,t} \sim \text{binomial}(n_{2,t-1}, \phi_2) \end{pmatrix}. \tag{2.2}$$

We will develop this example in subsequent sections, where it will be referred to as Example 1.

More generally, suppose we have m states (types of animal), with survival rates $\phi_1, \ldots, \phi_m$. Then the expected numbers of survivors may be expressed

$$\begin{bmatrix} \text{E}(u_{1(s),1,t}) \\ \text{E}(u_{1(s),2,t}) \\ \vdots \\ \text{E}(u_{1(s),m,t}) \end{bmatrix} = \begin{bmatrix} \phi_1 & 0 & \ldots & 0 \\ 0 & \phi_2 & \ldots & 0 \\ \vdots & \vdots & \ddots & \vdots \\ 0 & 0 & \ldots & \phi_m \end{bmatrix} \begin{bmatrix} n_{1,t-1} \\ n_{2,t-1} \\ \vdots \\ n_{m,t-1} \end{bmatrix}$$

with $u_{1(s),j,t} \sim \text{binomial}(n_{j,t-1}, \phi_j)$ for $j = 1, \ldots, m$.

Available data may not support m distinct survival parameters. A simple solution is to explore models in which some of the survival parameters are set equal; for example, a common adult survival rate is often assumed. We explore more flexible solutions, in which survival is modelled as a function of covariates or as a random effect, in Sect. 2.2.

In our matrix formulations, wherever we see $\text{E}(\cdot)$ on the left-hand side, indicating expectation of a random variable, we will need to specify a probability distribution for the corresponding stochastic process. If there is no expectation, then the corresponding process is deterministic. We now consider the deterministic process of age incrementation.

2.1.2 Age Incrementation

In Example 1, there are just two age classes. Assuming young animals become adults by the end of their first year, we can represent this as a deterministic matrix model: all survivors are deemed to be adults by the time of the next breeding season. The number of animals after age incrementation at the end of year t, given the number of survivors, is known without error:

$$\begin{bmatrix} u_{2(a),2,t} \end{bmatrix} = \begin{bmatrix} 1 & 1 \end{bmatrix} \begin{bmatrix} u_{1(s),1,t} \\ u_{1(s),2,t} \end{bmatrix}. \tag{2.3}$$

Note that this can equivalently be written

$$\begin{bmatrix} 0 \\ u_{2(a),2,t} \end{bmatrix} = \begin{bmatrix} 0 & 0 \\ 1 & 1 \end{bmatrix} \begin{bmatrix} u_{1(s),1,t} \\ u_{1(s),2,t} \end{bmatrix}.$$

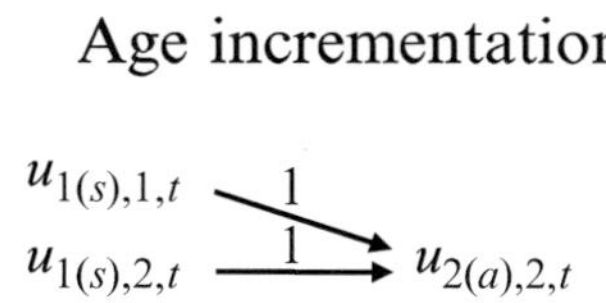

Fig. 2.2 Before age incrementation, we have $u_{1(s),1,t}$ young animals and $u_{1(s),2,t}$ adults in year t. After age incrementation, we have $u_{2(a),2,t} = u_{1(s),1,t} + u_{1(s),2,t}$ adults. The arrows are solid to indicate that the processes are deterministic, with $\mathrm{E}(u_{2(a),2,t}) \equiv u_{2(a),2,t}$. The rates associated with the processes are both unity because all animals follow the routes indicated by the arrows, and no new animals are generated

We adopt the convention that the dimension of the state vector is reduced when one or more states necessarily have no animals; the vector is then expanded again when a subsequent sub-process potentially generates animals that belong to these deleted states.

The effect of two ones in one row of a matrix is to combine two states; in this case, young and old animals. Equation (2.3) may be expressed diagrammatically as shown in Fig. 2.2. The model can equivalently be written $\mathbf{u}_{2(a),t} = \mathbf{A}\mathbf{u}_{1(s),t}$, where $\mathbf{A}$ is the age incrementation matrix. In fact, for this simple model and adopting the formulation of Eq. (2.3), $\mathbf{A}$ is a row vector, and $\mathbf{u}_{2(a),t}$ is a scalar.

If we have m states corresponding to ages 0, 1, 2, ..., then state m comprises all animals aged $m - 1$ or older. The following (deterministic) model ensures that each age class except the oldest moves up one year, and the two oldest age classes merge, at the year end. (The first entry of the state vector becomes $u_{2(a),2,t}$ because conceptually there are no animals in their first year until births occur.)

$$\begin{bmatrix} u_{2(a),2,t} \\ u_{2(a),3,t} \\ \vdots \\ u_{2(a),m-1,t} \\ u_{2(a),m,t} \end{bmatrix} = \begin{bmatrix} 1\,0 \ldots 0\,0\,0 \\ 0\,1 \ldots 0\,0\,0 \\ \vdots\ \vdots\ \ddots\ \vdots\ \vdots\ \vdots \\ 0\,0 \ldots 1\,0\,0 \\ 0\,0 \ldots 0\,1\,1 \end{bmatrix} \begin{bmatrix} u_{1(s),1,t} \\ u_{1(s),2,t} \\ \vdots \\ u_{1(s),m-2,t} \\ u_{1(s),m-1,t} \\ u_{1(s),m,t} \end{bmatrix}.$$

2.1.3 Growth

Suppose in Example 1, the two states are size classes rather than age classes, and suppose that at the year end, animals in the smaller size class move to the larger class with probability π. Animals in the larger class remain there. (The same model applies if the two classes correspond to immature and mature animals.) The expected numbers of animals in each state, conditional on the numbers before reallocation of animals according to their size, are then

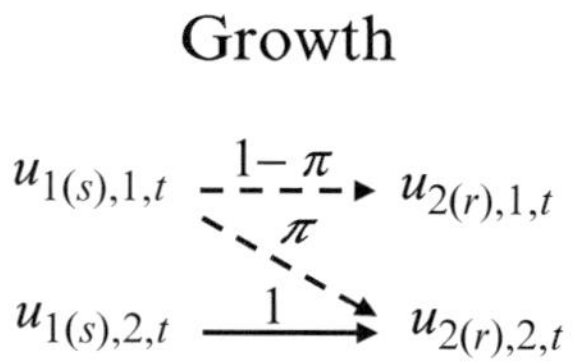

Fig. 2.3 Before growth, we have $u_{1(s),1,t}$ small animals and $u_{1(s),2,t}$ large animals in year t. After growth, we expect to have $\mathrm{E}(u_{2(r),1,t}) = (1-\pi)u_{1(s),1,t}$ small animals and $E(u_{2(r),2,t}) = \pi u_{1(s),1,t} + u_{1(s),2,t}$ large animals. The solid arrow with an associated rate of unity indicates that large animals remain large, while the dashed arrows indicate that the growth process for small animals is stochastic

$$\begin{bmatrix} \mathrm{E}(u_{2(r),1,t}) \\ \mathrm{E}(u_{2(r),2,t}) \end{bmatrix} = \begin{bmatrix} 1-\pi & 0 \\ \pi & 1 \end{bmatrix} \begin{bmatrix} u_{1(s),1,t} \\ u_{1(s),2,t} \end{bmatrix}, \tag{2.4}$$

which may also be expressed as $E(\mathbf{u}_{2(r),t}|\mathbf{u}_{1(s),t}) = \mathbf{R}\mathbf{u}_{1(s),t}$, where $\mathbf{R} = \begin{bmatrix} 1-\pi & 0 \\ \pi & 1 \end{bmatrix}$ is the growth matrix. This process is shown diagrammatically in Fig. 2.3.

As with survival, the binomial distribution is an appropriate model for stochasticity in the growth process:

$$\begin{pmatrix} u_{2(r),1,t} \sim \text{binomial}(u_{1(s),1,t}, 1-\pi) \\ u_{2(r),2,t} = u_{1(s),2,t} + (u_{1(s),1,t} - u_{2(r),1,t}) \end{pmatrix}. \tag{2.5}$$

If there are m size classes, and we allow animals in state j to remain the same with probability $1-\pi_j$ or move up one size class with probability π_j for states $1, 2, \ldots, m-1$, then

$$\begin{bmatrix} \mathrm{E}(u_{2(r),1,t}) \\ \mathrm{E}(u_{2(r),2,t}) \\ \mathrm{E}(u_{2(r),3,t}) \\ \vdots \\ \mathrm{E}(u_{2(r),m-1,t}) \\ \mathrm{E}(u_{2(r),m,t}) \end{bmatrix} = \begin{bmatrix} 1-\pi_1 & 0 & \ldots & 0 & 0 & 0 \\ \pi_1 & 1-\pi_2 & \ldots & 0 & 0 & 0 \\ 0 & \pi_2 & \ldots & 0 & 0 & 0 \\ \vdots & \vdots & \ddots & \vdots & \vdots & \vdots \\ 0 & 0 & \ldots & \pi_{m-2} & 1-\pi_{m-1} & 0 \\ 0 & 0 & \ldots & 0 & \pi_{m-1} & 1 \end{bmatrix} \begin{bmatrix} u_{1(s),1,t} \\ u_{1(s),2,t} \\ u_{1(s),3,t} \\ \vdots \\ u_{1(s),m-1,t} \\ u_{1(s),m,t} \end{bmatrix}.$$

Note that, if we allow the possibility that an animal in the smallest size class remains there ($\pi_1 < 1$), then we do not lose state 1 at this stage, in contrast with the aging model.

2.1.4 Birth

For Example 1, suppose births take place after age incrementation. Thus last year's young have been combined with adults, and births create the new cohort of young animals. If the birth rate is denoted by ρ, then

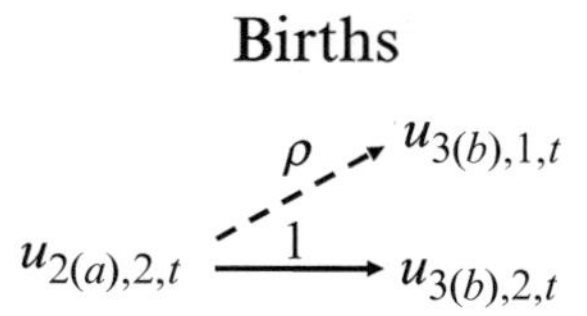

Fig. 2.4 Before births but after age incrementation, we have no young animals and $u_{2(a),2,t}$ adults in year t. After births, we expect to have $\mathrm{E}(u_{3(b),1,t}) = \rho u_{2(a),1,t}$ young animals, shown by the dashed arrow with associated rate ρ. The number of adults remains at $u_{3(b),2,t} = u_{2(a),2,t}$, indicated by a solid arrow with an associated rate of unity. The rates emanating from $u_{2(a),2,t}$ sum to $1 + \rho > 1$, indicating that new animals have been created

$$\begin{bmatrix} \mathrm{E}(u_{3(b),1,t}) \\ u_{3(b),2,t} \end{bmatrix} = \begin{bmatrix} \rho \\ 1 \end{bmatrix} \begin{bmatrix} u_{2(a),2,t} \end{bmatrix},$$

or equivalently, $E(\mathbf{u}_{3(b),t}|\mathbf{u}_{2(a),t}) = \mathbf{B}\mathbf{u}_{2(a),t}$ where $\mathbf{B} = \begin{bmatrix} \rho \\ 1 \end{bmatrix}$ is the birth matrix. (In this simple example, it is a column vector.) We illustrate this process in Fig. 2.4.

If each animal gives birth to at most one young per year, we might again invoke the binomial distribution:

$$\begin{pmatrix} u_{3(b),1,t} \sim \text{binomial}(u_{2(a),2,t}, \rho) \\ u_{3(b),2,t} = u_{2(a),2,t} \end{pmatrix}. \tag{2.6}$$

Thus the newly-born animals are placed in the recreated state for young animals, and the number of adults remains unchanged after the birth sub-process.

With m states, we have

$$\begin{bmatrix} \mathrm{E}(u_{3(b),1,t}) \\ u_{3(b),2,t} \\ u_{3(b),3,t} \\ \vdots \\ u_{3(b),m,t} \end{bmatrix} = \begin{bmatrix} \rho_2 & \rho_3 & \dots & \rho_m \\ 1 & 0 & \dots & 0 \\ 0 & 1 & \dots & 0 \\ \vdots & \vdots & \ddots & \vdots \\ 0 & 0 & \dots & 1 \end{bmatrix} \begin{bmatrix} u_{2(a),2,t} \\ u_{2(a),3,t} \\ \vdots \\ u_{2(a),m,t} \end{bmatrix}$$

where ρ_i is set to zero for states i that correspond to immatures or males.

To generalize the birth model, we need distributions that allow multiple births to a single mother. Possible distributions include the Poisson and negative binomial, but a more flexible choice is the multinomial distribution, where the number of trials is equal to the number of breeding females, and the probability that a female gives birth to i young is p_i for $i = 0, 1, \ldots$, with $\sum_i p_i = 1$ and $\rho = \sum_i i p_i$.

If the birth process completes the modelling of the annual cycle, then we have $u_{3(b),j,t} \equiv n_{j,t}$ for $j = 1, \ldots, m$.

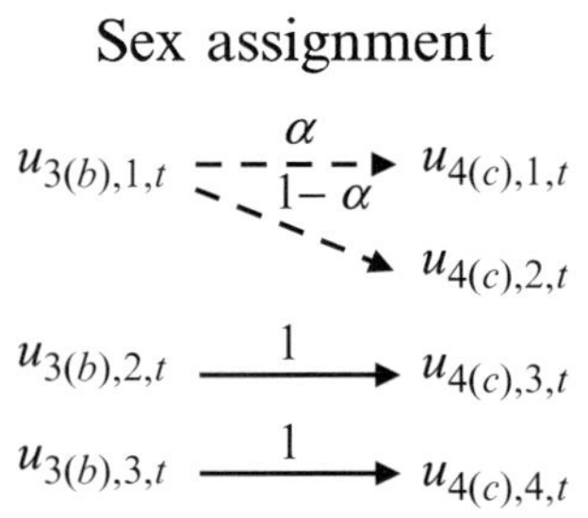

Fig. 2.5 Before sex assignment but after births, there are $u_{3(b),1,t}$ young animals in year t. We expect $\mathrm{E}(u_{4(c),1,t}) = \alpha u_{3(b),1,t}$ of these to be female, and the remainder male. For adult females, we have $u_{4(c),3,t} = u_{3(b),2,t}$, while for adult males, $u_{4(c),4,t} = u_{3(b),3,t}$

2.1.5 Sex Assignment

Commonly, only the female component of a population is modelled. Suppose we wish to add males to Example 1, and handle them as additional states. Thus we now have four states, corresponding to adult and young of each sex. We specify our sex assignment model conditional on number of births $u_{3(b),1,t}$. If $u_{3(b),2,t}$ is number of adult females, $u_{3(b),3,t}$ is number of adult males, and the probability that a young animal is female is α, then

$$\begin{bmatrix} \mathrm{E}(u_{4(c),1,t}) \\ \mathrm{E}(u_{4(c),2,t}) \\ u_{4(c),3,t} \\ u_{4(c),4,t} \end{bmatrix} = \begin{bmatrix} \alpha & 0\,0 \\ 1-\alpha & 0\,0 \\ 0 & 1\,0 \\ 0 & 0\,1 \end{bmatrix} \begin{bmatrix} u_{3(b),1,t} \\ u_{3(b),2,t} \\ u_{3(b),3,t} \end{bmatrix} \tag{2.7}$$

where $u_{4(c),1,t}$, $u_{4(c),2,t}$, $u_{4(c),3,t}$ and $u_{4(c),4,t}$ are numbers of newborn females, newborn males, adult females and adult males respectively. If this is the last process of the year, then $u_{4(c),j,t} \equiv n_{j,t}$ for $j = 1, 2, 3, 4$.

Equation (2.7) may be expressed as $E(\mathbf{u}_{4(c),t}|\mathbf{u}_{3(b),t}) = \mathbf{C}\mathbf{u}_{3(b),t}$, where $\mathbf{C}$ is the sex assignment matrix, and is represented by Fig. 2.5.

The binomial is the natural model for sex assignment:

$$\begin{pmatrix} u_{4(c),1,t} \sim \text{binomial}(u_{3(b),1,t}, \alpha) \\ u_{4(c),2,t} = u_{3(b),1,t} - u_{4(c),1,t} \\ u_{4(c),3,t} = u_{3(b),2,t} \\ u_{4(c),4,t} = u_{3(b),3,t} \end{pmatrix}.$$

If there are m age classes for each sex, then this generalizes straightforwardly:

$$\begin{bmatrix} E(u_{4(c),1,t}) \\ E(u_{4(c),2,t}) \\ u_{4(c),3,t} \\ u_{4(c),4,t} \\ \vdots \\ u_{4(c),2m,t} \end{bmatrix} = \begin{bmatrix} \alpha & 0 & 0 & \dots & 0 \\ 1-\alpha & 0 & 0 & \dots & 0 \\ 0 & 1 & 0 & \dots & 0 \\ 0 & 0 & 1 & \dots & 0 \\ \vdots & \vdots & \vdots & \ddots & \vdots \\ 0 & 0 & 0 & \dots & 1 \end{bmatrix} \begin{bmatrix} u_{3(b),1,t} \\ u_{3(b),2,t} \\ u_{3(b),3,t} \\ u_{3(b),4,t} \\ \vdots \\ u_{3(b),2m-1,t} \end{bmatrix}.$$

2.2 Models Within Models

In the previous section, models were presented for which all animals in a given state were subject to the same processes with the same parameter values, and those values were treated as fixed over time. We require flexible models with more parameters, yet we are unlikely to have adequate data to fit large numbers of parameters. We consider two solutions to this dilemma, each of which appreciably enhances the capabilities and realism of our models.

2.2.1 Modelling Processes Through Covariates

The first process model we considered was survival. We specified that animals in state j all had the same survival rate ϕ_j. However, even for animals in the same state, the survival rate may vary according to many different factors. Suppose we wish to model first-year survival ϕ_1 as a function of winter rainfall r_t and total abundance $N_{t-1} = \sum_{j=1}^{m} n_{j,t-1}$ at the end of year $t-1$. We would like our model to respect the constraint that ϕ_1 must lie between zero and one. A logistic model is a natural choice:

$$\phi_{1,t-1} = \frac{1}{1 + \exp\{\beta_0 + \beta_1 r_t + \beta_2 N_{t-1}\}}. \tag{2.8}$$

where $\phi_{1,t-1}$ is the probability that animals recruited into the population at the end of year $t-1$ survive to the end of year t. This incorporates an environmental effect, with $\beta_1 > 0$ if high winter rainfall increases first-year mortality, and density dependence, with $\beta_2 > 0$ if high abundance increases mortality. The above formulation corresponds to using a logit link function. Other link functions for binary data could also be used: the probit and the complementary log–log links (McCullagh and Nelder 1989:31).

This model allows ϕ_1 to be time-varying. We could also include covariates that vary by space (e.g. habitat) or by individual animal (e.g. weight). Thus individual-based models are feasible, without the necessity of more parameters than the data can support.

The rate parameters associated with other processes could be modelled similarly. For example the growth model parameter π of Sect. 2.1.3 might be modelled as a function of time spent in that size class and of resources available, or the birth rate ρ of Sect. 2.1.4 might be modelled as a function of the mother's weight, or a measure of her position in the social hierarchy, or of the habitat she occupies—or all three.

We note that some classic nonlinear, density-dependent population dynamic models such as the Beverton-Holt and Ricker models, which typically simultaneously incorporate survival and reproduction, can be extended to include covariates (Maunder and Deriso, 2011).

2.2.2 *Processes as Random Effects*

Often, we expect rates to vary but we do not have relevant covariates to model that variation. In this case, an option is to specify the rate as a random effect. For example, instead of specifying that the survival rate ϕ_j is the same for all animals in state j, we can assign a random distribution to ϕ_j, allowing it to vary in time, or by location, or by individual animal. A good candidate distribution is the beta distribution, which is constrained to lie between zero and one. Thus if we wish ϕ_j to vary by individual animal, instead of having to fit a different parameter for each animal (impossible in practice), we simply have to fit the two parameters of the beta distribution. For example, in the case of time-varying survival, for animals in state j at time t, the survival rate is $\phi_{j,t} \sim \text{beta}(\alpha_j, \beta_j)$. More commonly, random effects are assumed to be normal on some scale; in the case of modelling survival, the natural scale would be logit, ensuring that survival is constrained between zero and one. Such models as these where the parameters themselves are random variables are sometimes called hierarchical or multi-level models; random observations are at the lowest level and random parameters at higher levels. The parameters of the distributions at the higher levels are called hyperparameters.

Random effects and covariates can be combined in various ways to yield even more flexible models. For example, first-year survival in year t could be a function of a covariate, such as rainfall, and a random effect. Instead of Eq. (2.8), we might have

$$\phi_{1,t-1} = \frac{1}{1 + \exp(\beta_{0,t} + \beta_1 r_t)}$$

where

$$\beta_{0,t} \sim \text{normal}\left(\mu_0, \sigma^2\right).$$

This is an example of a nonlinear mixed effects model, where the effect of rainfall is a fixed effect and the intercept $\beta_{0,t}$ is a random effect.

Alternatively, the hyperparameters of a hierarchical model can be modelled as functions of covariates. Again referring to the example of first-year survival, one of the hyperparameters of the beta distribution could be a function of rainfall:

$$\phi_{1,t} \sim \text{beta}\left(\alpha(r_t), \beta_1\right)$$

where

$$\alpha(r_t) = \exp\left(\gamma_0 + \gamma_1 r_t\right).$$

2.3 Leslie Matrices and Lefkovitch Matrices

We now return to Example 1 with just two states, corresponding to the two age classes. We start the year just after the breeding season, so that the first process is survival, followed by age incrementation, and then births. Taking these in order, we have:

$$\begin{bmatrix} \mathrm{E}(u_{1(s),1,t}) \\ \mathrm{E}(u_{1(s),2,t}) \end{bmatrix} = \begin{bmatrix} \phi_1 & 0 \\ 0 & \phi_2 \end{bmatrix} \begin{bmatrix} n_{1,t-1} \\ n_{2,t-1} \end{bmatrix}$$

$$\begin{bmatrix} u_{2(a),2,t} \end{bmatrix} = \begin{bmatrix} 1 & 1 \end{bmatrix} \begin{bmatrix} u_{1(s),1,t} \\ u_{1(s),2,t} \end{bmatrix}$$

$$\begin{bmatrix} \mathrm{E}(n_{1,t}) \\ n_{2,t} \end{bmatrix} = \begin{bmatrix} \mathrm{E}(u_{3(b),1,t}) \\ u_{3(b),2,t} \end{bmatrix} = \begin{bmatrix} \rho \\ 1 \end{bmatrix} \begin{bmatrix} u_{2(a),2,t} \end{bmatrix}.$$

We may combine these into a single model simply by expressing the process matrices as a product in reverse chronological order:

$$\begin{bmatrix} \mathrm{E}(n_{1,t}) \\ \mathrm{E}(n_{2,t}) \end{bmatrix} = \begin{bmatrix} \rho \\ 1 \end{bmatrix} \begin{bmatrix} 1 & 1 \end{bmatrix} \begin{bmatrix} \phi_1 & 0 \\ 0 & \phi_2 \end{bmatrix} \begin{bmatrix} n_{1,t-1} \\ n_{2,t-1} \end{bmatrix}, \tag{2.9}$$

or alternatively, $E(\mathbf{n}_t|\mathbf{n}_{t-1}) = \mathbf{BASn}_{t-1}$. Evaluating the product, we obtain

$$\begin{bmatrix} \mathrm{E}(n_{1,t}) \\ \mathrm{E}(n_{2,t}) \end{bmatrix} = \begin{bmatrix} \rho\phi_1 & \rho\phi_2 \\ \phi_1 & \phi_2 \end{bmatrix} \begin{bmatrix} n_{1,t-1} \\ n_{2,t-1} \end{bmatrix}. \tag{2.10}$$

Provided that none of ϕ_1, ϕ_2 and ρ depend on numbers of animals present in year t, the expectations in the above expression are exact. For example Eq. (2.10) implies that $\mathrm{E}(n_{1,t}) = \rho\left(\phi_1 n_{1,t-1} + \phi_2 n_{2,t-1}\right)$. As the expectations are conditional

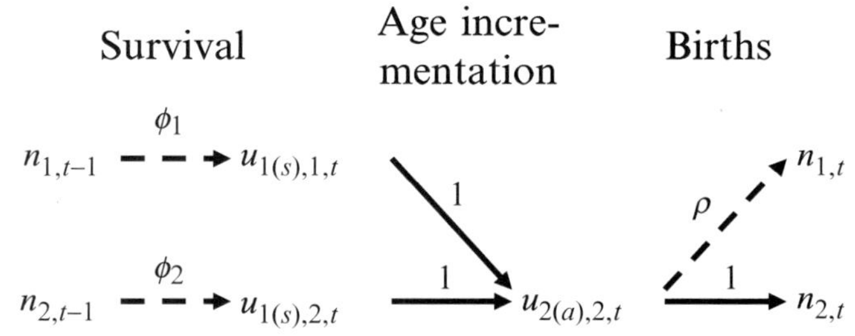

Fig. 2.6 In the BAS model, the sub-processes of Figs. 2.1, 2.2 and 2.4 are combined to create a model for which the annual processes, in chronological order, are survival (S), age incrementation (A) and births (B). Note that $n_{j,t} = u_{3(b),j,t}$ for $j = 1, 2$

on $n_{1,t-1}$ and $n_{2,t-1}$, then they hold if ϕ_1, ϕ_2 and ρ are either independent of numbers of animals in each state or dependent only on the states at time $t-1$ (or earlier). However, density-dependent effects in the birth rate might be more plausibly modelled by expressing ρ as a function of $n_{2,t}$, as this is the number of breeding adults present at the end of year t when births occur. In this case, the above expectation no longer holds.

The above matrix representations are useful for defining and understanding models. As will be seen in Chap. 4, we can fit these models without having to assume that expectations of the type shown in Eq. (2.10) hold. When such methods are used, the approximation is of no consequence.

The significance of Eq. (2.10) is that the product of the process matrices is an example of a Leslie matrix (Leslie 1945, 1948; Caswell 2001:8–11). (Note that the standard Leslie matrix would have ρ where we have $\rho\phi_j$; in our formulation, animals must survive to the year end to breed.) We call the above model a BAS model, because the Leslie matrix is obtained by taking the product of the matrices corresponding to the processes birth (B), age incrementation (A) and survival (S), in that order. The model is shown diagrammatically in Fig. 2.6.

If we replace the deterministic age incrementation model of Eq. (2.3) by the growth model of Eq. (2.4), and assume that only animals in the larger class breed, then we obtain the following Lefkovitch matrix (Lefkovitch 1965; Caswell 2001:59):

$$\begin{bmatrix} \mathrm{E}(n_{1,t}) \\ \mathrm{E}(n_{2,t}) \end{bmatrix} = \begin{bmatrix} 1 & \rho \\ 0 & 1 \end{bmatrix} \begin{bmatrix} 1-\pi & 0 \\ \pi & 1 \end{bmatrix} \begin{bmatrix} \phi_1 & 0 \\ 0 & \phi_2 \end{bmatrix} \begin{bmatrix} n_{1,t-1} \\ n_{2,t-1} \end{bmatrix}$$

$$= \begin{bmatrix} (1-\pi+\rho\pi)\phi_1 & \rho\phi_2 \\ \pi\phi_1 & \phi_2 \end{bmatrix} \begin{bmatrix} n_{1,t-1} \\ n_{2,t-1} \end{bmatrix}. \qquad (2.11)$$

Taking the three matrices in the first line of Eq. (2.11) in reverse order, the first gives us the expected numbers of animals that survive the year, given the numbers alive at the start of the year; the second gives the expected numbers of survivors

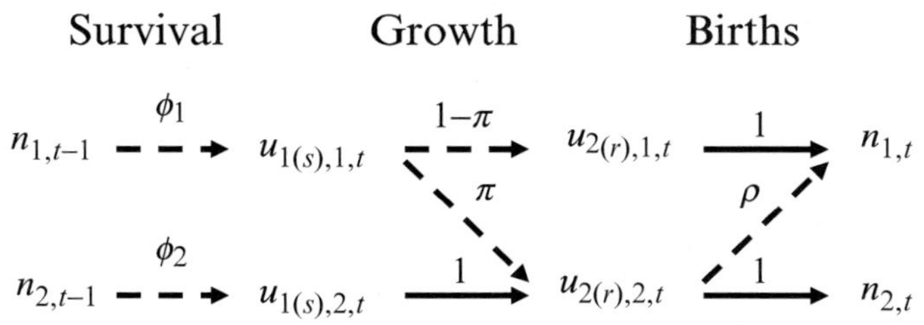

Fig. 2.7 In the BRS model, the sub-processes of Figs. 2.1, 2.3 and 2.4 are combined to create a model for which the annual processes, in chronological order, are survival (S), growth (R) and births (B). As in Fig. 2.6, $n_{j,t} = u_{3(b),j,t}$ for $j = 1, 2$

in the smaller size class that either remain in that size class or grow into the larger size class; and the third gives the expected number of newly-born animals that join the smaller size class. This is therefore a BRS model: the annual cycle starts with survival (S), then growth (R), and finally births (B) (Fig. 2.7). The Lefkovitch matrix of Eq. (2.11) is obtained as the matrix product **BRS**.

In general, if we take the product of the process matrices arranged in reverse chronological order, we obtain the *population projection matrix*[1]. Our modular approach allows for easy definition of more complex models. For example, if we retain the same processes as for the Leslie matrix of Eq. (2.10), i.e. survival, aging and births, but expand to include m states, then we obtain

$$\begin{bmatrix} E(n_{1,t}) \\ E(n_{2,t}) \\ \vdots \\ E(n_{m,t}) \end{bmatrix} = \begin{bmatrix} \rho_2 & \rho_3 & \dots & \rho_m \\ 1 & 0 & \dots & 0 \\ 0 & 1 & \dots & 0 \\ \vdots & \vdots & \ddots & \vdots \\ 0 & 0 & \dots & 1 \end{bmatrix} \begin{bmatrix} 1 & 0 & \dots & 0 & 0 & 0 \\ 0 & 1 & \dots & 0 & 0 & 0 \\ \vdots & \vdots & \ddots & \vdots & \vdots & \vdots \\ 0 & 0 & \dots & 1 & 0 & 0 \\ 0 & 0 & \dots & 0 & 1 & 1 \end{bmatrix} \begin{bmatrix} \phi_1 & 0 & \dots & 0 \\ 0 & \phi_2 & \dots & 0 \\ \vdots & \vdots & \ddots & \vdots \\ 0 & 0 & \dots & \phi_m \end{bmatrix} \begin{bmatrix} n_{1,t-1} \\ n_{2,t-1} \\ \vdots \\ n_{m,t-1} \end{bmatrix}$$

$$= \begin{bmatrix} \rho_2\phi_1 & \rho_3\phi_2 & \dots & \rho_m\phi_{m-1} & \rho_m\phi_m \\ \phi_1 & 0 & \dots & 0 & 0 \\ 0 & \phi_2 & \dots & 0 & 0 \\ \vdots & \vdots & \ddots & \vdots & \\ 0 & 0 & \dots & \phi_{m-1} & \phi_m \end{bmatrix} \begin{bmatrix} n_{1,t-1} \\ n_{2,t-1} \\ \vdots \\ n_{m,t-1} \end{bmatrix}.$$

This model is shown diagrammatically in Fig. 2.8.

The sex assignment model of Sect. 2.1.5, with different survival rates for males and females as well as for adults and young, yields

[1] Note that this matrix does not satisfy the definition of a projection matrix from linear algebra—the term "projection" is used here to indicate projection of the population from one time point to the next.

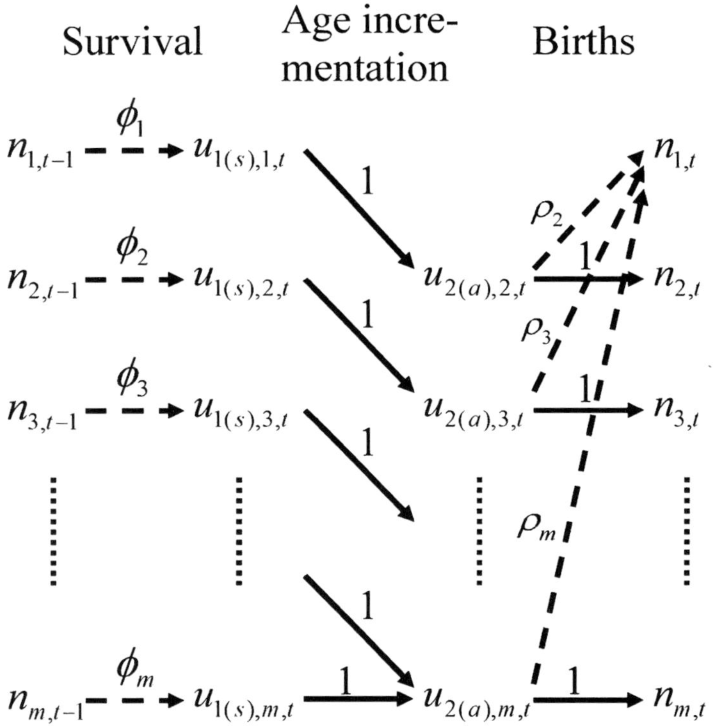

Fig. 2.8 The BAS model with m age classes. Compare this with Fig. 2.6

$$\begin{bmatrix} E(n_{1,t}) \\ E(n_{2,t}) \\ E(n_{3,t}) \\ E(n_{4,t}) \end{bmatrix} = \begin{bmatrix} \alpha & 0\,0 \\ 1-\alpha & 0\,0 \\ 0 & 1\,0 \\ 0 & 0\,1 \end{bmatrix} \begin{bmatrix} \rho\,0 \\ 1\,0 \\ 0\,1 \end{bmatrix} \begin{bmatrix} 1\,0\,1\,0 \\ 0\,1\,0\,1 \end{bmatrix} \begin{bmatrix} \phi_1 & 0 & 0 & 0 \\ 0 & \phi_2 & 0 & 0 \\ 0 & 0 & \phi_3 & 0 \\ 0 & 0 & 0 & \phi_4 \end{bmatrix} \begin{bmatrix} n_{1,t-1} \\ n_{2,t-1} \\ n_{3,t-1} \\ n_{4,t-1} \end{bmatrix}$$

$$= \begin{bmatrix} \alpha\rho\phi_1 & 0 & \alpha\rho\phi_3 & 0 \\ (1-\alpha)\rho\phi_1 & 0 & (1-\alpha)\rho\phi_3 & 0 \\ \phi_1 & 0 & \phi_3 & 0 \\ 0 & \phi_2 & 0 & \phi_4 \end{bmatrix} \begin{bmatrix} n_{1,t-1} \\ n_{2,t-1} \\ n_{3,t-1} \\ n_{4,t-1} \end{bmatrix}. \quad (2.12)$$

The four matrices that combine to form the generalized Leslie matrix of Eq. (2.12) correspond to the following sub-processes. First chronologically but last of the four matrices in the first line of Eq. (2.12) is the matrix that handles the survival process (S) of young and adult males and females. Next comes age incrementation (A), where young females merge with the adults, and similarly for males, temporarily resulting in just two states: adult females and adult males. The matrix for the birth process (B) comes next, generating a single state for newly-born animals.

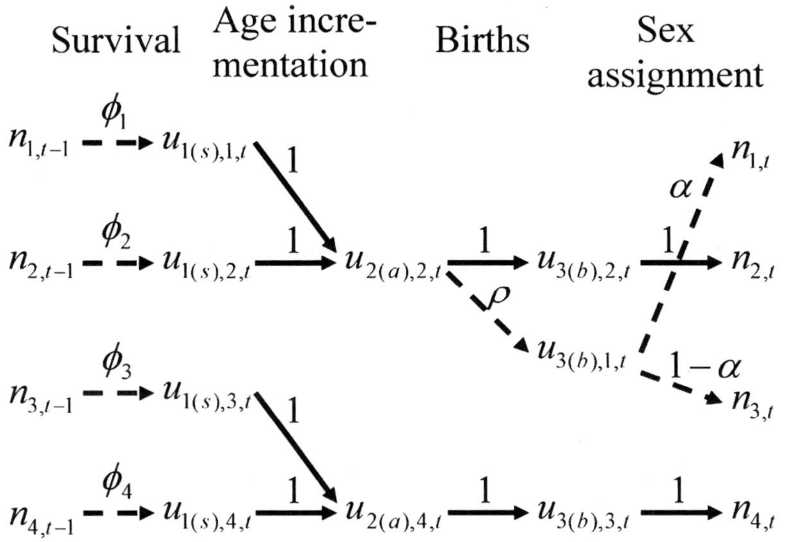

Fig. 2.9 The CBAS model: Example 1 with the addition of male animals. The *top row* corresponds to young females, then adult females, followed by young males and finally adult males. The *extra row* in the middle in the column for the birth sub-process corresponds to young before assignment of sex

Finally comes sex assignment (C), where this state is separated out into males and females. The matrix product **CBAS** therefore yields the generalized Leslie matrix. This CBAS model is illustrated in Fig. 2.9.

Even for such a simplistic model for population dynamics, it is starting to become apparent how much easier it is to formulate the model using this modular approach rather than direct specification of the population projection matrix. Caswell (2001:60) provides an example of how easily an error can occur in model formulation with the latter approach.

2.4 Using Matrices to Represent Processes of Multiple Populations

Here we consider situations where there are either subpopulations within a given species which are distinguished by criteria other than age or sex, or multiple populations of different species. In the former case we give two examples: one where the subpopulations are different genotypes of the same species and another where the subpopulations occupy different distinct locations, with some degree of movement between locations, and thus form a metapopulation. We could have included two-sex models in this section, but we chose to cover sex assignment within Sect. 2.1 because we consider that the state corresponding to the sex of an animal is one of the fundamental states in any population dynamics model, even if for many purposes, it is sufficient to model the female component of the population only.

We first consider how offspring may be assigned to genotype, so that the population is split into genotypes. We then show how to model movement between the components of a metapopulation. Next we develop multiple-species models, first to show how inter-species competition might be addressed, then to illustrate how predator-prey systems may be modelled using our framework.

2.4.1 Modelling Genotypes

Suppose animals are classified as to whether they are first-year or older (adult), whether they are male or female, and whether they have two dominant genes at a particular locus (DD), a dominant and recessive gene (Dd), or two recessive genes (dd). Suppose further that age incrementation is sub-process k in the annual cycle, so that just after age incrementation in year t, we have $u_{k(a),1,t}$ adult females of type DD, $u_{k(a),2,t}$ adult males of type DD, $u_{k(a),3,t}$ adult females of type Dd, $u_{k(a),4,t}$ adult males of type Dd, $u_{k(a),5,t}$ adult females of type dd, and $u_{k(a),6,t}$ adult males of type dd. We first consider the expected number of young born to each female genotype. For simplicity, we assume that all genotypes have the same birth rate. Denoting number of births to females of type DD by $u_{k+1(b),1,t}$, number of births to females of type Dd by $u_{k+1(b),2,t}$, and number of births to females of type dd by $u_{k+1(b),3,t}$, we have

$$\begin{bmatrix} E(u_{k+1(b),1,t}) \\ E(u_{k+1(b),2,t}) \\ E(u_{k+1(b),3,t}) \\ u_{k+1(b),4,t} \\ u_{k+1(b),5,t} \\ \vdots \\ u_{k+1(b),9,t} \end{bmatrix} = \begin{bmatrix} \rho & 0 & 0 & 0 & 0 & 0 \\ 0 & 0 & \rho & 0 & 0 & 0 \\ 0 & 0 & 0 & 0 & \rho & 0 \\ 1 & 0 & 0 & 0 & 0 & 0 \\ 0 & 1 & 0 & 0 & 0 & 0 \\ \vdots & \vdots & \vdots & \vdots & \vdots & \vdots \\ 0 & 0 & 0 & 0 & 0 & 1 \end{bmatrix} \begin{bmatrix} u_{k(a),1,t} \\ u_{k(a),2,t} \\ u_{k(a),3,t} \\ u_{k(a),4,t} \\ u_{k(a),5,t} \\ u_{k(a),6,t} \end{bmatrix}.$$

We now need to reallocate the births, not according to the genotype of the mother, but according to the genotype of the young. Denote the probability that the father of a young animal has genotype DD, given that the mother is of type DD, by $\psi_{DD|DD}$, the probability that the father has genotype Dd, given that the mother has genotype DD, by $\psi_{Dd|DD}$, and so on. Then we can express the probability that a young animal is of a given genotype, given the genotype of the mother, as follows: $\eta_{DD|DD} = \psi_{DD|DD} + \frac{1}{2}\psi_{Dd|DD}$; $\eta_{DD|Dd} = \frac{1}{2}\psi_{DD|Dd} + \frac{1}{4}\psi_{Dd|Dd}$; $\eta_{Dd|DD} = \frac{1}{2}\psi_{Dd|DD} + \psi_{dd|DD}$; $\eta_{Dd|Dd} = \frac{1}{2}(\psi_{DD|Dd} + \psi_{Dd|Dd} + \psi_{dd|Dd})$; $\eta_{Dd|dd} = \frac{1}{2}\psi_{Dd|dd} + \psi_{DD|dd}$; $\eta_{dd|Dd} = \frac{1}{4}\psi_{Dd|Dd} + \frac{1}{2}\psi_{dd|Dd}$; $\eta_{dd|dd} = \frac{1}{2}\psi_{Dd|dd} + \psi_{dd|dd}$. Let $u_{k+2(g),1,t}$ be the number of young of genotype DD, $u_{k+2(g),2,t}$ the number of young of genotype

Dd, $u_{k+2(g),3,t}$ the number of young of genotype dd, $u_{k+2(g),4,t}$ the number of adult females of genotype DD, and so on. Then conditional on the numbers of young born to each female genotype, expected numbers are:

$$\begin{bmatrix} \mathrm{E}(u_{k+2(g),1,t}) \\ \mathrm{E}(u_{k+2(g),2,t}) \\ \mathrm{E}(u_{k+2(g),3,t}) \\ u_{k+2(g),4,t} \\ u_{k+2(g),5,t} \\ \vdots \\ u_{k+2(g),9,t} \end{bmatrix} = \begin{bmatrix} \eta_{DD|DD} & \eta_{DD|Dd} & 0 & 0\,0 \ldots 0 \\ \eta_{Dd|DD} & \eta_{Dd|Dd} & \eta_{Dd|dd} & 0\,0 \ldots 0 \\ 0 & \eta_{dd|Dd} & \eta_{dd|dd} & 0\,0 \ldots 0 \\ 0 & 0 & 0 & 1\,0 \ldots 0 \\ 0 & 0 & 0 & 0\,1 \ldots 0 \\ \vdots & \vdots & \vdots & \vdots\,\vdots \ddots \vdots \\ 0 & 0 & 0 & 0\,0 \ldots 1 \end{bmatrix} \begin{bmatrix} u_{k+1(b),1,t} \\ u_{k+1(b),2,t} \\ u_{k+1(b),3,t} \\ u_{k+1(b),4,t} \\ u_{k+1(b),5,t} \\ \vdots \\ u_{k+1(b),9,t} \end{bmatrix}.$$

Assuming sex is independent of genotype, we can now assign sex to the young by extending the method of Sect. 2.1.5 appropriately.

Conditional on the $u_{k+1(b),j,t}$, we can specify the distributions of the $u_{k+2(g),j,t}$ as:

$$\begin{pmatrix} u_{k+2(g),1,t} = r_{DD,DD} + r_{DD,Dd} \\ u_{k+2(g),2,t} = r_{Dd,DD} + r_{Dd,Dd} + r_{Dd,dd} \\ u_{k+2(g),3,t} = r_{Dd,dd} + r_{dd,dd} \\ u_{k+2(g),4,t} = u_{k+1(b),4,t} \\ \vdots \\ u_{k+2(g),9,t} = u_{k+1(b),9,t} \end{pmatrix}$$

with

$$\begin{pmatrix} r_{DD,DD} \sim \text{binomial}(u_{k+1(b),1,t}, \eta_{DD|DD}) \\ r_{Dd,DD} = u_{k+1(b),1,t} - r_{DD,DD} \\ (r_{DD,Dd}, r_{Dd,Dd}, r_{dd,Dd}) \sim \text{multinomial}(u_{k+1(b),2,t}, \eta_{DD|Dd}, \eta_{Dd|Dd}, \eta_{dd|Dd}) \\ r_{Dd,dd} \sim \text{binomial}(u_{k+1(b),3,t}, \eta_{Dd|dd}) \\ r_{dd,dd} = u_{k+1(b),3,t} - r_{Dd,dd} \end{pmatrix}.$$

Note that $r_{Dd,DD}$ for example denotes the number of offspring from genotype DD females that are of genotype Dd.

Under a random mating model, we have $\psi_{DD|DD} = \psi_{DD|Dd} = \psi_{DD|dd} = u_{k(a),2,t}/(u_{k(a),2,t} + u_{k(a),4,t} + u_{k(a),6,t})$ and similarly for other probabilities. More strictly, these proportions should relate to when the animals mate. For example if they mate a year before births occur, then $\psi_{DD|DD}$ in year t should be $u_{k(a),2,t-1}/(u_{k(a),2,t-1} + u_{k(a),4,t-1} + u_{k(a),6,t-1})$.

This approach is readily extendible to more complex genetics models. However, it would be advisable to develop computer algorithms for the tedious task of model formulation! Even this simple model is too complex to illustrate diagrammatically, due to the large number of arrows that would be required—especially if matings were represented in the diagram.

2.4.2 Modelling Metapopulations

Returning to Example 1 with just two age classes (see Sect. 2.1.1), suppose our population occupies two distinct sites. Suppose further that movement occurs after winter survival, but before the breeding season. Let $u_{1(s),1,t}$ be the number of surviving young and $u_{1(s),2,t}$ number of surviving adults in site 1 in year t, and let $u_{1(s),3,t}$ and $u_{1(s),4,t}$ be the young and adult survivors in site 2. Let $\mu_{1\to 2}$ be the probability that an animal in site 1 moves to site 2, assumed for simplicity to be the same for young and adults, and similarly $\mu_{2\to 1}$ is the probability of movement from site 2 to site 1. If $u_{2(m),1,t}$ represents number of young in site 1 after movement, with corresponding definitions for adults in site 1 ($u_{2(m),2,t}$), and young ($u_{2(m),3,t}$) and adults ($u_{2(m),4,t}$) in site 2, then expected numbers after movement, given numbers before movement, are

$$\begin{bmatrix} \mathrm{E}(u_{2(m),1,t}) \\ \mathrm{E}(u_{2(m),2,t}) \\ \mathrm{E}(u_{2(m),3,t}) \\ \mathrm{E}(u_{2(m),4,t}) \end{bmatrix} = \begin{bmatrix} 1-\mu_{1\to 2} & 0 & \mu_{2\to 1} & 0 \\ 0 & 1-\mu_{1\to 2} & 0 & \mu_{2\to 1} \\ \mu_{1\to 2} & 0 & 1-\mu_{2\to 1} & 0 \\ 0 & \mu_{1\to 2} & 0 & 1-\mu_{2\to 1} \end{bmatrix} \begin{bmatrix} u_{1(s),1,t} \\ u_{1(s),2,t} \\ u_{1(s),3,t} \\ u_{1(s),4,t} \end{bmatrix}. \quad (2.13)$$

Assuming binomial distributions, we have

$$\begin{pmatrix} u_{2(m),1,t} = w_{3,t} + (u_{1(s),1,t} - w_{1,t}) \\ u_{2(m),2,t} = w_{4,t} + (u_{1(s),2,t} - w_{2,t}) \\ u_{2(m),3,t} = w_{1,t} + (u_{1(s),3,t} - w_{3,t}) \\ u_{2(m),4,t} = w_{2,t} + (u_{1(s),4,t} - w_{4,t}) \end{pmatrix}$$

where

$$\begin{pmatrix} w_{1,t} \sim \mathrm{binomial}(u_{1(s),1,t}, \mu_{1\to 2}) \\ w_{2,t} \sim \mathrm{binomial}(u_{1(s),2,t}, \mu_{1\to 2}) \\ w_{3,t} \sim \mathrm{binomial}(u_{1(s),3,t}, \mu_{2\to 1}) \\ w_{4,t} \sim \mathrm{binomial}(u_{1(s),4,t}, \mu_{2\to 1}) \end{pmatrix}$$

Putting all the sub-processes together, namely survival (S), movement (M), age incrementation (A) and birth (B), we have a BAMS model:

$$\begin{bmatrix} \mathrm{E}(n_{1,t}) \\ \mathrm{E}(n_{2,t}) \\ \mathrm{E}(n_{3,t}) \\ \mathrm{E}(n_{4,t}) \end{bmatrix} = \begin{bmatrix} \rho & 0 \\ 1 & 0 \\ 0 & \rho \\ 0 & 1 \end{bmatrix} \begin{bmatrix} 1 & 1 & 0 & 0 \\ 0 & 0 & 1 & 1 \end{bmatrix} \begin{bmatrix} 1-\mu_{1\to 2} & 0 & \mu_{2\to 1} & 0 \\ 0 & 1-\mu_{1\to 2} & 0 & \mu_{2\to 1} \\ \mu_{1\to 2} & 0 & 1-\mu_{2\to 1} & 0 \\ 0 & \mu_{1\to 2} & 0 & 1-\mu_{2\to 1} \end{bmatrix}$$

$$\times \begin{bmatrix} \phi_1 & 0 & 0 & 0 \\ 0 & \phi_2 & 0 & 0 \\ 0 & 0 & \phi_3 & 0 \\ 0 & 0 & 0 & \phi_4 \end{bmatrix} \begin{bmatrix} n_{1,t-1} \\ n_{2,t-1} \\ n_{3,t-1} \\ n_{4,t-1} \end{bmatrix}. \quad (2.14)$$

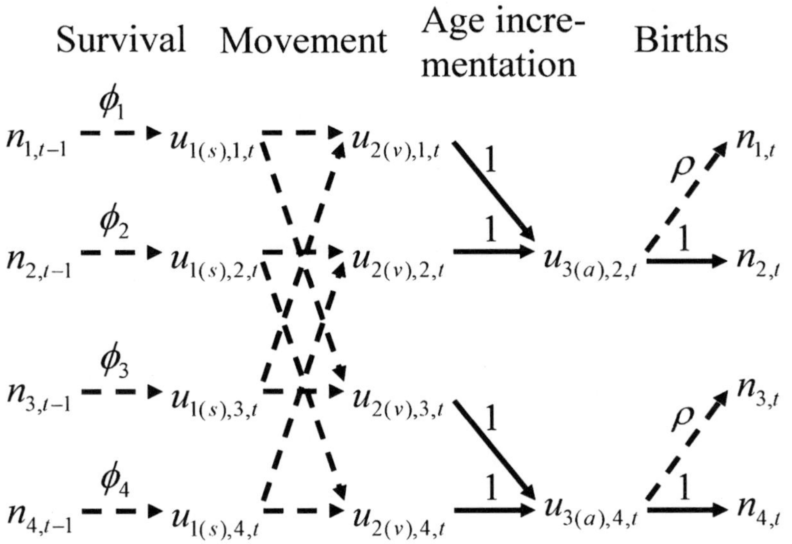

Fig. 2.10 The BAMS model: Example 1 but with two sites, and movement between them. The top row corresponds to young in site 1, then adults in site 1, followed by young in site 2 and finally adults in site 2. In this formulation, a different survival rate is assumed for each of the four states, but the birth rate ρ is assumed to be the same in each site. Within a site, young and adults have the same movement rate, but this is allowed to differ between sites. Rates associated with movement arrows are not plotted for clarity. The top two horizontal arrows have rate $1 - \mu_{1\to2}$, the bottom two horizontal arrows have rate $1 - \mu_{2\to1}$, the arrows from site 1 (top) to site 2 have rate $\mu_{1\to2}$, and the arrows from site 2 to site 1 have rate $\mu_{2\to1}$

Starting with the last of the projection matrices in Eq. (2.14), survival of the adults and young in each site is the first process of the year. The next matrix handles movement; diagonal terms correspond to animals that stay in their current site, while off-diagonal terms correspond to those that move. Age incrementation follows, in which the young animals at each site are merged with the adults, leaving just two states. The birth matrix then recreates four states, by generating newly-born animals in each site. The processes of this BAMS model are illustrated in Fig. 2.10.

Note that the movement sub-process matrix in Eq. (2.13) can be split into submatrices:

$$\begin{bmatrix} 1-\mu_{1\to2} & 0 & \mu_{2\to1} & 0 \\ 0 & 1-\mu_{1\to2} & 0 & \mu_{2\to1} \\ \mu_{1\to2} & 0 & 1-\mu_{2\to1} & 0 \\ 0 & \mu_{1\to2} & 0 & 1-\mu_{2\to1} \end{bmatrix} = \begin{bmatrix} M_{1\to1} & M_{2\to1} \\ M_{1\to2} & M_{2\to2} \end{bmatrix}$$

where $M_{1\to 1} = \begin{bmatrix} 1-\mu_{1\to 2} & 0 \\ 0 & 1-\mu_{1\to 2} \end{bmatrix}$ corresponds to animals that stay in site 1, $M_{2\to 1} = \begin{bmatrix} \mu_{2\to 1} & 0 \\ 0 & \mu_{2\to 1} \end{bmatrix}$ corresponds to animals that move from site 2 to site 1, and so on. Thus a general movement model is given by

$$\mathrm{E}(\mathbf{u}_{2(m),t}) = \begin{bmatrix} M_{1\to 1} & M_{2\to 1} & \ldots & M_{l\to 1} \\ M_{1\to 2} & M_{2\to 2} & \ldots & M_{l\to 2} \\ \vdots & \vdots & \ddots & \vdots \\ M_{1\to l} & M_{2\to l} & \ldots & M_{l\to l} \end{bmatrix} \mathbf{u}_{1(s),t}$$

where the vector $\mathbf{u}_{1(s),t}$ has elements equal to numbers of animals in each state before movement, and $\mathbf{u}_{2(m),t}$ represents numbers of animals by state after movement. For the case of just two states (age classes) per site, submatrix $M_{i\to j}$ is given by

$$M_{i\to j} = \begin{bmatrix} \mu_{i\to j} & 0 \\ 0 & \mu_{i\to j} \end{bmatrix}$$

for $i, j = 1, \ldots, l$, where $\mu_{i\to i} = 1 - \sum_{j\neq i} \mu_{i\to j}$ is the probability that an animal in site i remains there. For more states per site, these submatrices expand in the obvious way.

Given l sites, the numbers of animals at site i at time t that remain there or move to one of the other sites can be modelled by a multinomial

$$(u_{i\to 1}, \ldots, u_{i\to l}) \sim \text{multinomial}\,(\mu_{i\to 1}, \ldots, \mu_{i\to l})$$

where $\sum_{j=1}^{l} \mu_{i\to j}=1$. With the constraint on the sum of probabilities, there are $l-1$ movement parameters. Thus with l sites there are $l \times (l-1)$ parameters in total. Unless large-scale mark and release studies are carried out across all sites, empirical model-free estimates of $\mu_{i\to j}$ will generally not be available.

In the absence of individual site-to-site movement data, the use of covariates to model the movement parameters, as described in Sect. 2.2.1, is a pragmatic alternative. Site-specific information such as distances to other sites, animal densities at each site, and measures of site habitat quality are potential covariates. For example, Thomas et al. (2005) modelled the probability of movement between British grey seal colonies as a function of distance, animal density and site fidelity. Modelling must be constrained to ensure that the probabilities are non-negative and sum to one, but this can be easily handled by using exponential functions and then rescaling appropriately. Suppose one covariate x_1 is used. Then, for example,

$$p_{i\to j} = \exp\left(\beta_0 + \beta_1 x_{1i} + \beta_2 x_{1j}\right)$$

$$\mu_{i\to j} = \frac{p_{i\to j}}{\sum_l p_{i\to l}}.$$

Hierarchical or random effects models can be used, in combination with any relevant covariates. Now (see Sect. 2.2.2) the movement parameters $\mu_{i \to j}$ are themselves random variables. A natural distribution for proportions that must sum to one is the Dirichlet distribution (a multivariate extension of the beta distribution). Similar to an example given in Buckland et al. (2004), suppose that there are three colonies in a metapopulation and that colony densities and distances between colonies affect the probability of movement. Consider movement from colony 2, for example:

$$(\mu_{2\to 1}, \mu_{2\to 2}, \mu_{2\to 3}) \sim \text{Dirichlet}\,(\lambda_{2\to 1}, \lambda_{2\to 2}, \lambda_{2\to 3})$$

with

$$\begin{aligned}\lambda_{2\to 2} &= \exp(\beta_0)\\ \lambda_{2\to j} &= \exp\left(\beta_0 + \beta_1 N_{2t} + \beta_2 H_{2t} + \beta_3 N_{jt} + \beta_4 H_{jt} + \beta_5 d_{2j}\right), \quad j = 1, 3,\end{aligned}$$

where $\beta_1 > 0$, $\beta_2 < 0$, $\beta_3 < 0$, $\beta_4 > 0$ and $\beta_5 < 0$; N_{it} is the abundance at site i in year t, H_{it} is a measure of habitat quality with larger values indicating greater suitability, and d_{ij} is the distance between sites i and j. Given the parameters $\mu_{i \to j}$, movement is then modelled by a multinomial distribution.

2.4.3 *Multi-Species Models*

We can readily extend our models to multi-species systems by modelling rates for one species as functions of abundance of other species. For example, a competition model for a two-species system might make the survival or birth rate of one species a decreasing function of the abundance of the other. A predator-prey model might make the survival rate of the prey a decreasing function of the abundance of the predator, and the survival and birth rates of the predator an increasing function of the abundance of the prey. A host-parasite model might be similarly formulated. More complex interactions might be modelled for community models. For example, a model for marine fish stocks might allow adults of one species to prey on first-year fish of another, but to be prey for adults of the other species. It might also incorporate competition effects between species with similar diets, and movement rates might be modelled as functions of abundance of each species at different locations.

As an example of a multi-species model, we consider a system comprising one predator and two prey, with competition between the two prey species. Suppose we model the predator P in two age classes, first-year animals and adults. For simplicity, we model the two prey species A and B each as a single state. First-year survival of predators is assumed to depend on abundance of both prey species, whereas adult survival depends on abundance of species B alone. Species A is assumed to suffer a competition effect from species B, but not *vice versa*. All species show density dependence in their survival rates, but for the predators, this affects first-year survival alone. Then a simple model to describe this system is as follows.

Denote number of first-year predators, adult predators, prey species A and prey species B at the end of year t by $n_{0,t}$, $n_{1,t}$, $n_{A,t}$ and $n_{B,t}$ respectively. Assuming that mortality (S) first occurs, followed by age incrementation (A), and then births (B), we have a multi-species BAS model:

$$\begin{bmatrix} \mathrm{E}(n_{0,t}) \\ \mathrm{E}(n_{1,t}) \\ \mathrm{E}(n_{A,t}) \\ \mathrm{E}(n_{B,t}) \end{bmatrix} = \begin{bmatrix} \rho_P & 0 & 0 \\ 1 & 0 & 0 \\ 0 & 1+\rho_A & 0 \\ 0 & 0 & 1+\rho_B \end{bmatrix} \begin{bmatrix} 1\,1\,0\,0 \\ 0\,0\,1\,0 \\ 0\,0\,0\,1 \end{bmatrix}$$

$$\times \begin{bmatrix} \phi_{0,t-1} & 0 & 0 & 0 \\ 0 & \phi_{1,t-1} & 0 & 0 \\ 0 & 0 & \phi_{A,t-1} & 0 \\ 0 & 0 & 0 & \phi_{B,t-1} \end{bmatrix} \begin{bmatrix} n_{0,t-1} \\ n_{1,t-1} \\ n_{A,t-1} \\ n_{B,t-1} \end{bmatrix}$$

where ρ_P, ρ_A and ρ_B are the birth rates of predators, prey species A and prey species B respectively, and the survival parameters ϕ are defined as

$$\phi_{0,t-1} = \frac{1}{1+\exp\{\alpha_1 - \beta_1 n_{A,t-1} - \beta_2 n_{B,t-1} + \beta_3(n_{0,t-1} + n_{1,t-1})\}}$$

$$\phi_{1,t-1} = \frac{1}{1+\exp(\alpha_2 - \beta_4 n_{B,t-1})}$$

$$\phi_{A,t-1} = \frac{1}{1+\exp\{\alpha_3 + \beta_5 n_{A,t-1} + \beta_6 n_{B,t-1} + \beta_7(n_{0,t-1} + n_{1,t-1})\}}$$

$$\phi_{B,t-1} = \frac{1}{1+\exp\{\alpha_4 + \beta_8 n_{B,t-1} + \beta_9(n_{0,t-1} + n_{1,t-1})\}}$$

where the α and β are parameters (which may be estimated using the methods of Chap. 4), with each $\beta > 0$, and where for example $\phi_{A,t-1}$ is the probability that an animal of species A that is alive at the end of year $t-1$ survives to the end of year t. Note that the birth rates may similarly be modelled, for example to allow lower birth rates at high densities of a given species.

2.5 Beyond Matrix Models

Matrices are linear operators. Consider the BAS model defined by Eq. (2.9). We may rewrite this equation

$$\mathbf{n}_t = \tilde{L}_{t-1}(\mathbf{n}_{t-1})$$

where the *process operator* $\tilde{L}_{t-1}(\cdot)$ is the composition $\tilde{L}_{t-1}(\cdot) = \tilde{L}_{b,t}(\tilde{L}_{a,t}(\tilde{L}_{s,t-1}(\cdot)))$, and $\tilde{L}_{s,t-1}(\cdot)$, $\tilde{L}_{a,t}(\cdot)$ and $\tilde{L}_{b,t}(\cdot)$ are the survival, aging and birth operators respectively (Buckland et al. 2007). Note that, for consistency with notation elsewhere in this book, we have adopted a different convention from Buckland et al. (2007) in that we give the operator $\tilde{L}_{t-1}$ the same suffix as the state vector $\mathbf{n}_{t-1}$ on the right-hand side of the above equation, as the operator acts on $\mathbf{n}_{t-1}$, whereas Buckland et al. (2007) defined $\mathbf{n}_t = \tilde{L}_t(\mathbf{n}_{t-1})$, giving the operator the suffix corresponding to the state vector on the left-hand side. Similarly, we give the survival operator $\tilde{L}_{s,t-1}$ suffix $t-1$, but the aging and birth operators act on animals that have survived to the end of year t, so have suffixes t.

If all sub-processes are deterministic, we obtain the classical matrix model

$$\mathbf{n}_t = \mathbf{L}_{t-1}\mathbf{n}_{t-1},$$

where $\mathbf{L}_{t-1} = \mathbf{BAS}$ is typically a generalized Leslie or Lefkovitch matrix. Note that elements of $\mathbf{L}_{t-1}$ can depend on $\mathbf{n}_{t-1}$, in which case the model is nonlinear in the states. Thus for example density dependence can be modelled in this framework.

The special case of most interest here occurs when one or more of the sub-processes is stochastic, but the expected values of the elements of $\mathbf{n}_t$ may be expressed as functions of the elements of $\mathbf{n}_{t-1}$. Then

$$E(\mathbf{n}_t|\mathbf{n}_{t-1}) = \mathbf{L}_{t-1}\mathbf{n}_{t-1}, \qquad (2.15)$$

where $\mathbf{L}_{t-1}$ is a population projection matrix such that $E(\tilde{L}_{t-1}(\mathbf{n}_{t-1})|\mathbf{n}_{t-1}) = \mathbf{L}_{t-1}\mathbf{n}_{t-1}$. The BAS model of Eq. (2.9) is an example of this, in which $\mathbf{L}_{t-1}$ is the Leslie matrix of Eq. (2.10).

If we take the model of Sect. 2.4.3, but model the predator birth rate ρ_P as a function of prey and/or predator abundance immediately preceding the births, then this yields a more general example, for which $E(\tilde{L}_{t-1}(\mathbf{n}_{t-1})|\mathbf{n}_{t-1}) \neq \mathbf{L}_{t-1}\mathbf{n}_{t-1}$, although if the nonlinearity is not strong, we might expect the result to hold approximately. The model fitting algorithms of Chap. 4 do not need Eq. (2.15) to hold.

2.6 Observation Matrices

For practical wildlife management, the ability to fit a population matrix model to a time series of data is important. To use the model fitting algorithms of Chap. 4, we need to specify distributions for observations that respect their relationship with the (usually unobserved) states in $\mathbf{n}_t$.

As in Sect. 2.5 for the process operator, we can define a general random operator for the observation process:

$$\mathbf{y}_t = \tilde{\mathbf{O}}_t(\mathbf{n}_t)$$

where $\mathbf{y}_t$ is a vector of observations for year t (Buckland et al. 2007). If the operator is linear, then we obtain the following observation equation:

$$\mathbf{y}_t = \mathbf{O}_t \mathbf{n}_t + \boldsymbol{\epsilon}_t$$

for the appropriate matrix $\mathbf{O}_t$. Assuming $E(\boldsymbol{\epsilon}_t|\mathbf{n}_t) = \mathbf{0}$, then

$$E(\mathbf{y}_t|\mathbf{n}_t) = \mathbf{O}_t \mathbf{n}_t. \tag{2.16}$$

In this formulation, we assume that the observations are taken at the end of year t. If this is not the case, then the state vector $\mathbf{n}_t$ in the above equation would be replaced by the appropriate intermediate state vector. For example, if population size was estimated after winter mortality occurs but before breeding starts, then the intermediate state vector would correspond to the survivors from the previous year.

As an example of an observation matrix, consider the BAMS model of Sect. 2.4.2. Suppose we have estimates $y_{1,t}$ and $y_{2,t}$ of total population size in site 1 and site 2 respectively at the end of year t (just after breeding). Suppose further that we assume that these estimates are independently normally distributed with variances σ_1^2 and σ_2^2 respectively. Then

$$\begin{bmatrix} y_{1,t} \\ y_{2,t} \end{bmatrix} = \begin{bmatrix} 1 & 1 & 0 & 0 \\ 0 & 0 & 1 & 1 \end{bmatrix} \begin{bmatrix} n_{1,t} \\ n_{2,t} \\ n_{3,t} \\ n_{4,t} \end{bmatrix} + \boldsymbol{\epsilon}_t \tag{2.17}$$

with $\boldsymbol{\epsilon}_t = \begin{bmatrix} \epsilon_{1,t} \\ \epsilon_{2,t} \end{bmatrix} \sim \text{normal}\left(\begin{bmatrix} 0 \\ 0 \end{bmatrix}, \begin{bmatrix} \sigma_1^2 & 0 \\ 0 & \sigma_2^2 \end{bmatrix} \right)$.

If the estimates of abundance are of adults only, then the observation matrix becomes $\mathbf{O}_t = \begin{bmatrix} 0 & 1 & 0 & 0 \\ 0 & 0 & 0 & 1 \end{bmatrix}$.

Usually, the observational study in year t would provide estimates of σ_1^2 and σ_2^2. Often those estimates are treated as known values when fitting a state-space model to data as simultaneously estimating variances for the state process and for the observation model is often difficult due to identifiability problems (Dennis et al. 2006). An alternative, when using the Bayesian methods of Chap. 4, is to use these estimates to inform the prior distributions for σ_1^2 and σ_2^2 (although such double usage of the data is not strictly Bayesian).

The observations in the above example are single point estimates of the individual states along with estimates of the variances. Multiple point estimates of the individual states are easily incorporated by inserting additional components in the observation vector and duplicating rows in the observation matrix.

A less trivial variation in the observation model is to work with the *raw* sample data, the measurements actually made in the field say, rather than summaries of the sample data. Typically point estimates of state vector components will be summaries of the actual measurements taken. For example, mark-recapture studies might be carried out over a sequence of years and estimates of population abundance constructed from the data. Rather than treat the point estimates as observations, the mark recoveries could be the observations. This is an intuitively attractive idea in that all estimation is done in a single step and complete within-sample variation is retained. A practical limitation, in some situations, is that the raw data may be unavailable or difficult to access. A technical limitation is that, to use the raw data in a matrix model, the observations must be written as linear functions of the states. The ease with which such a linear mapping can be done will depend upon how the data were collected, and may in some cases, e.g. distance sampling, be at best cumbersome and at worst infeasible.

2.7 Defining a Population Dynamics Model

We consider two examples to illustrate the steps involved in formulating a matrix model for population dynamics. In the first, we develop a model for coho salmon and in the second a metapopulation of deer, comprising just two populations with movement between them. We will describe the dynamics in deterministic or expected value terms and in the next chapter we will develop a fully stochastic formulation.

2.7.1 Coho Salmon

The following description of a matrix model for coho salmon *Oncorhynchus kisutch* (Fig. 2.11) is based on a model described in Newman (1998). That model fitted recoveries (in samples taken from ocean fishery catches) of marked coho salmon released from a hatchery located on a river on the west coast of the Olympic Peninsula in Washington State USA. The population sub-processes included survival, movement in the ocean, harvest, and migration back to the natal river.

Juvenile coho salmon are bred and reared in a hatchery for approximately 18 months. They are then released from the hatchery in May to enter the river where they travel downstream to enter the Pacific Ocean. Shortly before release, the fish are marked (with a batch-specific mark). Upon entry to the ocean they disperse up and down the coast and experience natural mortality (from predators, disease, etc). About 14 months after entry (roughly July of the following year), they begin to be harvested by ocean fisheries in 12 different regions (management areas) along the coast. The ocean fishery operates for up to 16 weeks and then the surviving, and now mature adult fish, migrate back to the river where they are harvested in the river, return to the hatchery, or spawn in the river and die.

The data include the number of marked fish released from the hatchery (denoted R), the number of marked fish caught in the ocean fisheries, stratified

Fig. 2.11 Leaping coho salmon. Photo: courtesy of Northwest Indian Fisheries Commission

by management area and week ($y_{a,t}$ for area a, $a = 1, \ldots, 12$ and week t, $t = 1, \ldots, 16$), and the number of fish returning to the river as mature adults (y_T).

The matrix model is initialized by calculating a vector of abundances in the management areas prior to the fishing season, i.e. the initial state vector. First the expected total number of survivors (summed over all regions) from the R released fish to the beginning of the fishing season is determined, i.e. $R\phi_J$, where ϕ_J is the survival probability from time of release to beginning of fishing. The expected numbers per area are allocated using the probabilities for a beta(α_i, β_i) distribution, its domain being a line segment beginning at the southern boundary of the most southern fishing region to the northern boundary of the most northern fishing region, i.e. the values from the beta distribution are essentially latitudinal coordinates. The fishing regions are thus defined as a non-overlapping (and exhaustive) partitioning of the line segment into 12 individual segments. The expected initial abundance in area a is then $n_{a,0} = R\phi_J p_a$, where p_a is the probability of being present in area a.

The heart of the matrix model is the projection of the abundances by fishing management area on a week by week basis during the fishing season. The vector of abundances is denoted $\mathbf{n}'_t = [n_{1,t}, n_{2,t}, \ldots, n_{12,t}]$. At the beginning of each week, mortality occurs, a combination of natural and harvest mortality, and movement follows. The survival probability in a given area is $\phi_{a,t} = \exp(-N - F_{a,t})$, where N is the natural mortality rate parameter, assumed constant, and $F_{a,t}$ is the fishing mortality rate parameter (and is a function of the fishing effort in the area that week). Movement from one area to another is a function of location and time (details in

Newman 1998) and the probability of moving from area a to area b during week t is denoted $m_{a\to b,t}$. The expected abundances in week t are written as:

$$E[\mathbf{n}_t|\mathbf{n}_{t-1}] = \mathbf{M}_t\mathbf{S}_{t-1}\mathbf{n}_{t-1} \tag{2.18}$$

where

$$\mathbf{S}_{t-1} = \begin{bmatrix} \phi_{1,t-1} & 0 & \dots & 0 \\ 0 & \phi_{2,t-1} & \dots & 0 \\ \vdots & \vdots & \dots & \vdots \\ 0 & 0 & \dots & \phi_{12,t-1} \end{bmatrix} \tag{2.19}$$

and

$$\mathbf{M}_t = \begin{bmatrix} m_{1\to 1,t} & m_{2\to 1,t} & \dots & m_{12\to 1,t} \\ m_{1\to 2,t} & m_{2\to 2,t} & \dots & m_{12\to 2,t} \\ \vdots & \vdots & \dots & \vdots \\ m_{1\to 12,t} & m_{2\to 12,t} & \dots & m_{12\to 12,t} \end{bmatrix}. \tag{2.20}$$

The catch data, $\mathbf{y}_t$, are linked to the state vector of abundances, $\mathbf{n}_t$, by a harvest matrix, $\mathbf{H}_t$. The elements $h_{a,t}$ of the harvest matrix are the fractions of mortality attributed to fishing:

$$h_{a,t} = \frac{F_{a,t}}{M + F_{a,t}}\left(1 - \exp(-M - F_{a,t})\right). \tag{2.21}$$

The expectation of the observations in matrix form is

$$E[\mathbf{y}_t|\mathbf{n}_t] = \mathbf{H}_t\mathbf{n}_t \tag{2.22}$$

where

$$\mathbf{H}_t = \begin{bmatrix} h_{1,t} & 0 & \dots & 0 \\ 0 & h_{2,t} & \dots & 0 \\ \vdots & \vdots & \dots & \vdots \\ 0 & 0 & \dots & h_{12,t} \end{bmatrix}. \tag{2.23}$$

Let P_I be a column vector of length 12 with elements representing probabilities of the initial locations at the beginning of the fishing season. Then the matrix models for the states and observations are summarized below.

$$E[\mathbf{n}_0] = R\phi_J P_I \tag{2.24}$$

$$E[\mathbf{n}_t|\mathbf{n}_{t-1}] = \mathbf{M}_t\mathbf{S}_{t-1}\mathbf{n}_{t-1}, \quad t = 1, \dots, 16 \tag{2.25}$$

$$E[\mathbf{y}_t|\mathbf{n}_t] = \mathbf{H}_t\mathbf{n}_t, \quad t = 1, \dots, 16. \tag{2.26}$$

Fig. 2.12 Red deer in Scotland provide an example of where state-space models have been used to help wildlife managers set cull levels (Trenkel et al. 2000). Photo: Steve Buckland

2.7.2 A Deer Metapopulation

The following example of two deer populations, labeled A and B, is hypothetical but it identifies some of the sub-processes, and their complexity, that underlie the dynamics of real metapopulations. It is based loosely on red deer *Cervus elaphus* dynamics (Fig. 2.12). As said previously, deterministic formulations are given here, with stochasticity added in the next chapter.

We arbitrarily take the year to start just after breeding, assumed to be early summer. Survival rate for calves is assumed to be density-dependent and a function of day-degrees frost in winter (a measure of winter severity equal to number of degrees of frost, summed over all days for which the daily low fell below zero), while adult survival (age one or more) is assumed to differ by sex, but to be constant otherwise. Movement between populations is assumed to occur just before breeding: conceptually, it is the population to which animals belong at breeding that we are primarily concerned with, rather than an accurate model of when animals switch between populations. Movement rate is assumed to be a function of animal density in the respective populations just before births occur. Age incrementation occurs after movement. One-year-old animals are assumed to have a separate, low birth rate, while older animals all have the same birth rate. Each female is assumed to give birth to at most one young. New-born animals can of course be male or female.

We now know all the processes which we need to model, but we have not fully specified the model for each process. To incorporate all the features of the conceptual model, we need just two age classes for males and three for females (to allow one-year-old females to have a lower birth rate), giving five states for each of two populations, and the state vector, $\mathbf{n}_t$, then has ten components. (If we wished to model survival or birth rates as more complex functions of age, we would need to retain more age classes.)

Consider first survival. We can express the expected number of survivors to year t from just after breeding at the end of year $t-1$ as $E(\mathbf{u}_{1(s),t}|\mathbf{n}_{t-1}) = \mathbf{S}_{t-1}\mathbf{n}_{t-1} = \begin{bmatrix} \mathbf{S}_{A,t-1} & \mathbf{0}_{5\times 5} \\ \mathbf{0}_{5\times 5} & \mathbf{S}_{B,t-1} \end{bmatrix} \mathbf{n}_{t-1}$, where $\mathbf{0}_{5\times 5}$ is a 5×5 matrix of zeros, and the survival matrices for populations A and B are

$$\mathbf{S}_{A,t-1} = \begin{bmatrix} \phi_{1,A,t-1} & 0 & 0 & 0 & 0 \\ 0 & \phi_{1,A,t-1} & 0 & 0 & 0 \\ 0 & 0 & \phi_f & 0 & 0 \\ 0 & 0 & 0 & \phi_f & 0 \\ 0 & 0 & 0 & 0 & \phi_m \end{bmatrix} \tag{2.27}$$

and

$$\mathbf{S}_{B,t-1} = \begin{bmatrix} \phi_{1,B,t-1} & 0 & 0 & 0 & 0 \\ 0 & \phi_{1,B,t-1} & 0 & 0 & 0 \\ 0 & 0 & \phi_f & 0 & 0 \\ 0 & 0 & 0 & \phi_f & 0 \\ 0 & 0 & 0 & 0 & \phi_m \end{bmatrix} \tag{2.28}$$

respectively. We have ordered the state vector so that the first element corresponds to first-year females in population A, followed by first-year males, second-year females, older females, and finally males in their second year or older. This is then repeated for population B.

The probabilities of the survival matrix, $\mathbf{S}_{t-1}$, vary by age class and sex. Survival for first year animals is assumed to be affected by both the population abundance (hence density dependent) as well as winter temperatures. We assume a logistic form for this dependence:

$$\phi_{1,A,t-1} = \frac{1}{1 + \exp\{\beta_0 + \beta_1 f_{A,t} + \beta_2 N_{A,t-1}\}} \tag{2.29}$$

where $f_{A,t}$ is number of day-degrees frost experienced by population A in year t, and $N_{A,t-1} = \sum_i n_{i,t-1}$ where summation is over the five states corresponding to population A, and so represents size of population A just after breeding. A similar model may be specified for $\phi_{1,B,t-1}$. Adult survival differs between sexes, with ϕ_f and ϕ_m the survival probabilities for females and males, respectively, but they are the same for both populations and are assumed constant over time. Thus the components of the survival matrix $\mathbf{S}_{t-1}$ that are time-dependent relate to survival of male and female first-year animals only.

The next model component required relates to movement, which we wish to be density-dependent. Denote the probability that an animal moves from population A to population B in year t by $\mu_{A\to B,t}$, and the probability that an animal moves from population B to population A by $\mu_{B\to A,t}$. The expected state vector after movement has occurred is then given by $E(\mathbf{u}_{2(m),t}|\mathbf{u}_{1(s),t}) = \mathbf{M}_t\mathbf{u}_{1(s),t} = \begin{bmatrix} \mathbf{M}_{A,A,t} & \mathbf{M}_{B,A,t} \\ \mathbf{M}_{A,B,t} & \mathbf{M}_{B,B,t} \end{bmatrix}\mathbf{u}_{1(s),t}$, where $\mathbf{M}_{A,A,t}$ is a 5×5 diagonal matrix with diagonal elements all equal to $1 - \mu_{A\to B,t}$, $\mathbf{M}_{A,B,t}$ is a 5×5 diagonal matrix with diagonal elements all equal to $\mu_{A\to B,t}$, and similarly for $\mathbf{M}_{B,B,t}$ and $\mathbf{M}_{B,A,t}$. The probability of movement is assumed to be a function of the difference in densities for the two populations,

$$\mu_{A\to B,t} = \frac{1}{1 + \exp\{\gamma_0 + \gamma_1(D_{A,t} - D_{B,t})\}} \tag{2.30}$$

where $D_{A,t} = \sum_i u_{i,1(s),t}/A_A$ is the density of animals in population A just before movement occurs, expressed as the total number of survivors, $\sum_i u_{i,1(s),t}$, where summation is over the five states for animals in population A, divided by the size A_A of the area occupied by population A. We define $D_{B,t}$ similarly. A model for $\mu_{B\to A,t}$ may be defined in the same way.

The next process is age incrementation. Thus we have $E(\mathbf{u}_{3(a),t}|\mathbf{u}_{2(m),t}) = \mathbf{A}\mathbf{u}_{2(m),t} = \begin{bmatrix} \mathbf{A}_A & \mathbf{0}_{3\times 5} \\ \mathbf{0}_{3\times 5} & \mathbf{A}_B \end{bmatrix}\mathbf{u}_{2(m),t}$ where $\mathbf{0}_{3\times 5}$ is a 3×5 matrix of zeros, and

$$\mathbf{A}_A = \mathbf{A}_B = \begin{bmatrix} 1 & 0 & 0 & 0 & 0 \\ 0 & 0 & 1 & 1 & 0 \\ 0 & 1 & 0 & 0 & 1 \end{bmatrix}. \tag{2.31}$$

Thus, the state vector now has three elements for each population, corresponding to second-year females, older females, and males in their second year or older.

We now define a model for generating new births. Only females can give birth, and one-year-old females (just entering their second year) have a lower birth rate than older females. Thus we have $E(\mathbf{u}_{4(b),t}|\mathbf{u}_{3(a),t}) = \mathbf{B}\mathbf{u}_{3(a),t} = \begin{bmatrix} \mathbf{B}_A & \mathbf{0}_{4\times 3} \\ \mathbf{0}_{4\times 3} & \mathbf{B}_B \end{bmatrix}\mathbf{u}_{3(a),t}$ where $\mathbf{0}_{4\times 3}$ is a 4×3 matrix of zeros, and

$$\mathbf{B}_A = \mathbf{B}_B = \begin{bmatrix} \rho_1 & \rho_2 & 0 \\ 1 & 0 & 0 \\ 0 & 1 & 0 \\ 0 & 0 & 1 \end{bmatrix}. \tag{2.32}$$

Thus ρ_1 is the probability that a one-year-old female gives birth, while ρ_2 is the probability that an older female gives birth.

Finally, we need a process model for assigning sex to the new-born animals. Suppose the probability that a new-born animal is female is α. If we considered that it was reasonable to assume that $\alpha = 0.5$, we would just fix it at this value. Otherwise, we can retain it as an unknown parameter to be estimated. In the latter case, we would like to include some data in the likelihood on observed proportion of females for a representative set of births. We can now write $E(\mathbf{n}_t|\mathbf{u}_{4(b),t}) \equiv E(\mathbf{u}_{5(c),t}|\mathbf{u}_{4(b),t}) = \mathbf{C}\mathbf{u}_{4(b),t} = \begin{bmatrix} \mathbf{C}_A & \mathbf{0}_{5\times4} \\ \mathbf{0}_{5\times4} & \mathbf{C}_B \end{bmatrix} \mathbf{u}_{4(b),t}$ where $\mathbf{0}_{5\times4}$ is a 5×4 matrix of zeros, and

$$\mathbf{C}_A = \mathbf{C}_B = \begin{bmatrix} \alpha & 0\,0\,0 \\ 1-\alpha & 0\,0\,0 \\ 0 & 1\,0\,0 \\ 0 & 0\,1\,0 \\ 0 & 0\,0\,1 \end{bmatrix}. \tag{2.33}$$

We now have a fully specified (deterministic) population dynamics model, but we have not yet addressed how the states relate to our data. Suppose we have annual estimates of number of adults (one-year-old or older) generated from aerial surveys of each population at the time of breeding. There is no information on age or sex. Then the expected observation equation for year t is

$$E[\mathbf{y}_t|\mathbf{n}_t] = \mathbf{O}\mathbf{n}_t \tag{2.34}$$

where $\mathbf{y}_t$ is a vector of length two, corresponding to an estimate of adult abundance in population A and population B, and

$$\mathbf{O} = \begin{bmatrix} 0\,0\,1\,1\,1\,0\,0\,0\,0\,0 \\ 0\,0\,0\,0\,0\,0\,0\,1\,1\,1 \end{bmatrix}. \tag{2.35}$$

The complete (deterministic) model for the population dynamics and the observation process is summarized as follows.

$$E[\mathbf{n}_t|\mathbf{n}_{t-1}] = \mathbf{L}_{t-1}\mathbf{n}_{t-1} = \mathbf{CBAM}_t\mathbf{S}_{t-1}\mathbf{n}_{t-1} \tag{2.36}$$

$$E[\mathbf{y}_t|\mathbf{n}_t] = \mathbf{O}\mathbf{n}_t. \tag{2.37}$$

The expected abundance at time t was written as a Leslie matrix operation, conditional on the previous abundance, i.e. $\mathbf{L}_{t-1}\mathbf{n}_{t-1}$. A look at each of the five component matrices indicates the complexity of the resulting matrix $\mathbf{L}_{t-1}$, a matrix that is relatively easily constructed by thinking in terms of sequential sub-processes, but near impossible to construct otherwise.

Chapter 3
State-Space Models

In the previous chapter, a sequence of matrices was used to model the sequence of subprocesses, birth, survival, movement, etc., which characterize population dynamics. We find this building block perspective attractive for at least two reasons: (1) it allows one to mentally "divide and conquer" sometimes complicated population dynamics processes; (2) the resulting product of matrices is a generalization of Leslie and Lefkovitch matrices, something familiar to many biologists.

While the building block matrix model is an aid to model formulation and understanding, it may not be as useful for fitting models to data and making population projections. Matrix representations of population dynamics describe at best the expected changes in a population, for example, conditional expected numbers at time t given numbers at time $t-1$, and often, as noted in Sect. 2.5, are just approximations of these expectations, as in the case of density-dependent birth processes. Consequently, such matrix models fail to describe the variation and uncertainty around these expected outcomes. There are various ways of extending matrix models to incorporate both variability (Quinn and Deriso 1999: Sect. 7.3; Caswell 2001: Chaps. 14 and 15) and nonlinearity (Quinn and Deriso 1999: Sect. 7.4; Caswell 2001: Chap. 16). For example, variability around the expected outcome can be described by randomly selecting elements of a matrix (Caswell 2001: Sect. 14.5.5) or randomly selecting a matrix from a set of matrices (Caswell 2001: Sect. 14.5.3).

However, limitations of stochastic and nonlinear matrix models become apparent when one considers modelling population dynamics that involve a sequence of random and nonlinear sub-processes. Example 1 from the previous chapter, the BAS model, included binomial distributions for survival of two different age classes and a third binomial distribution for births. A Leslie matrix was formulated that accurately characterised the expected states at time t (conditional on $\mathbf{n}_{t-1}$). For projecting the population forward in time, however, one cannot readily, if at all, formulate a stochastic version of the matrix that accurately captures the variation of these binomials. For example, simply adding a vector of random variables to the

K.B. Newman et al., *Modelling Population Dynamics: Model Formulation, Fitting and Assessment using State-Space Methods*, Methods in Statistical Ecology,
DOI 10.1007/978-1-4939-0977-3_3,

matrix model, i.e. $\mathbf{n}_t = BAS\mathbf{n}_{t-1} + \boldsymbol{\epsilon}_t$, is problematic at best. The distribution for $\boldsymbol{\epsilon}_t$ will be quite complex; e.g. its components cannot be so large that a corresponding component of $\mathbf{n}_t$ is less than 0.

State-space models (SSMs) are a more flexible approach for realistically modelling population dynamics than matrix models. SSMs are the backbone of the methods discussed in this book and this chapter is an introduction to their basic structure. Similar to some of the matrix models discussed in the previous chapter, SSMs have a state model and an observation model, but each is now a *stochastic time series*. The state model is a first-order Markov process, i.e. the distribution for state $\mathbf{n}_t$ is defined conditional on the previous state $\mathbf{n}_{t-1}$, and the distribution for the observation $\mathbf{y}_t$ is defined conditional on the current state $\mathbf{n}_t$. The classic SSM is a Normal Dynamic Linear Model (NDLM, West and Harrison 1997), which consists of two normally distributed linear models conditioning on $\mathbf{n}_{t-1}$ and $\mathbf{n}_t$, respectively. The two equations below are a simplistic example of an NDLM in the context of animal abundance dynamics, where N_t is the true, but unknown, abundance of an animal population at time t and y_t is an index of N_t, i.e. $y_t = \gamma N_t$ where γ is a constant of proportionality.

$$\text{State process model } N_t | N_{t-1} \sim \text{normal}\left(\lambda N_{t-1}, \sigma_N^2\right). \tag{3.1}$$

$$\text{Observation model } y_t | N_t \sim \text{normal}\left(\gamma N_t, \sigma_y^2\right). \tag{3.2}$$

Here λ is the population growth rate; in a deterministic setting, $\lambda > 1$ leads to exponential growth, and $\lambda < 1$ is exponential decline. We use the terms state process model and state model interchangeably. Variation around the expected value, in this case λN_{t-1}, is sometimes called process noise or variation. In this example, the magnitude of the process variation depends on the size of σ_N^2. An example of simulated projections of states with $\lambda = 1.02$ (2 % growth rate) and $\sigma_N^2 = 4$, and unbiased observations ($\gamma = 1$) with $\sigma_y^2 = 16$, is shown in Fig. 3.1.

Thus SSMs simultaneously account for two distinct sources of variation, natural or process variation (e.g. environmental or demographic stochasticity) and observation error (e.g. sampling or measurement errors) within a single framework. SSMs are much more general and flexible than matrix models, readily accommodating multiple random nonlinear sub-processes. Given the conditionally-defined state process model, forward stochastic population projection is relatively simple so long as random samples can be generated from the distribution. When the state process is a sequence of stochastic sub-processes, simulation is often easier than evaluation of the pdf; i.e. it is easier to simulate $\mathbf{n}_t$ given $\mathbf{n}_{t-1}$ than it is to calculate $\Pr(\mathbf{n}_t | \mathbf{n}_{t-1})$. The inclusion of a stochastic observation process model for the observations provides a framework for estimating parameters and accounting for uncertainty in the data, in a way which is consistent with the underlying state process model. This is in contrast to standard usage of matrix models where vital rate parameters, or their estimates, are somehow supplied external to the model, and error in such estimates is often not accounted for.

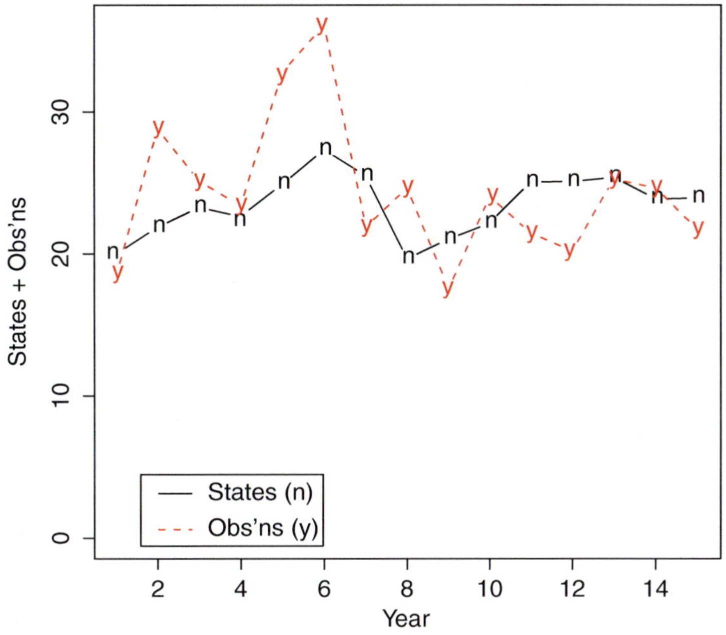

Fig. 3.1 Simulation of NDLM for animal abundance and estimates, where the population dynamics of the states are $N_t \sim \text{normal}(1.02N_{t-1}, 4)$ and estimates are unbiased, $y_t \sim \text{normal}(N_t, 16)$

In addition to the integrative nature of SSMs, there is also opportunity for a convenient division of labour between the modelling of the population dynamics and the modelling of the sampling and measurement of the population. Subject matter specialists such as ecologists can focus their attention on the underlying science through the state process model. The building block matrix models can serve as useful first approximations to the formulation of state process models, e.g. characterizing the deterministic portion of the state process model. Alternative and competing hypotheses about the underlying dynamics can be formalized by alternative state process models. Specialists in sampling, or mark-recapture, or transect sampling, or however the population is monitored, can focus on the formulation of one or more observation models. Different formulations of the observation model might result by working with summarized or derived calculations of sample data, e.g. mark-recapture-based point estimates of abundance, or the raw sample data, e.g. recaptured marks. This potential division of labour, of course, does not preclude a single individual, knowledgeable about the subject matter science and quantification of the observation process, from doing both.

Applications of state-space models to ecological data sets have steadily increased since the late 1980s. Early applications were largely restricted to the special case of NDLMs, because such models could be readily fitted using the Kalman filter (West and Harrison 1997). Several of the first applications were to fisheries data including

Mendelssohn (1988) who fitted an NDLM to Pacific mackerel *Scomber japonicus* catch data categorized by age class to estimate annual abundances and recruitment to the population, and Sullivan (1992) who fitted an NDLM to fisheries catch data categorized by length classes and then estimated parameters such as growth and survival. Advancements in computing power combined with simulation-based estimation procedures, such as Markov chain Monte Carlo (MCMC), extended the class of SSMs that could be fitted to nonlinear and non-normal distributions. Meyer and Millar (1999) gave one of the earliest demonstrations of such models for ecological data, South Atlantic albacore *Thunnus alalunga* biomass, using MCMC to fit an SSM where the state model was a univariate nonlinear, non-normal Schaefer surplus production model and the observation model used catch-per-unit-effort indices.

While many of the first applications of SSMs were to fisheries data, the diversity of species modelled by SSMs has expanded considerably. Animal species modelled by SSMs include red deer *Cervus elaphus* (Trenkel et al. 2000), grey herons *Ardea cinerea* and northern lapwings *Vanellus vanellus* (Besbeas et al. 2002), grey seals *Halichoerus grypus* (Thomas et al. 2005; Harrison et al. 2006), Chinook salmon *Oncorhynchus tshawytscha* (Newman and Lindley 2006), leatherback turtles *Dermochelys coriacea* (Jonsen et al. 2006), black bears *Ursus americanus* (Conn et al. 2008), red grouse *Lagopus lagopus scoticus* (New et al. 2009), hen harriers *Circus cyaneus* (New et al. 2011), desert bighorn sheep *Ovis canadensis mexicana* (Colchero et al. 2009), skate *Leucoraja ocellata* (Swain et al. 2009), Weddell seals *Leptonychotes weddelli* (Rotella et al. 2009), California sea lions *Zalophus californianus* (Ward et al. 2010), octopus *Octopus vulgaris* (Robert et al. 2010) and the Glanville fritillary butterfly *Melitaea cinxia* (Harrison et al. 2011).

State variables other than abundances have been considered as well. Royle and Kéry (2007) let the true state be whether or not a particular site was occupied by animals (and the observations were imperfect estimates of presence or absence). Gimenez et al. (2007) and Royle (2008), considering marked and recaptured or recovered animals, let the true state of marked animals (whether still alive or not) be the state variable. Anderson-Sprecher and Ledolter (1991) modelled the true location of radio-collared mule deer *Odoccoileus hemionus*, while Jonsen et al. (2006) did the same for individual tagged leatherback turtles. King (2014) describes how various types of ecological data can be modelled using a state-space formulation.

The biological processes explicitly modelled by the state model include mortality, birth or recruitment, and movement. Models for mortality have distinguished natural mortality for different life history stages (e.g. age 2, 3 and 4 Chinook salmon, Newman and Lindley 2006) and harvest-related mortality (e.g. black bears, Conn et al. 2008). Models for movement have included the movement of single animals (e.g. turtles, Jonsen et al. 2006), movement between areas of members of a single population (e.g. coho salmon, Newman 1998), and movement between metapopulations (e.g. four sets of pupping colonies used by grey seals, Thomas et al. 2005).

The remainder of this chapter includes a general statistical formulation of SSMs and simple examples. General approaches to inference, i.e. the fitting of SSMs to data, are discussed in Chap. 4, while Chap. 5 is a discussion of issues in SSM formulation and diagnostics. The remaining chapters show how to put specific problems, such as survival estimation (Chap. 7), into an SSM framework and include more detailed examples.

3.1 State-Space Models

Here we give a more formal and mathematically general definition of state-space models. State-space models are models for two discrete time processes running in parallel, one called the state process and the other the observation process. The state process is modelled by a conditional probability density function (pdf) or probability mass function (pmf) that describes the change of the state vector from time $t-1$ to time t, and is denoted by g_t. As will become clear in some of the later examples, g_t can be quite complicated and analytically intractable, reflecting a sequence of stochastic sub-processes. With some abuse of terminology, we will refer to pdf's and pmf's simply as pdf's. Note that we allow only discrete time indices, and to simplify notation, we assume that these are evenly spaced.

The value of the state vector at a given point in time is seldom known with certainty. Additionally, the parameters, such as survival rates and birth rates, are almost never known with certainty. If inferences about the state model are to be possible and defensible, field data must be collected. For example, various components of the state vector might be estimated from mark-recapture or line transect surveys. This leads to the observation process, which is modelled by another conditional pdf, denoted f_t, that describes the relationship between the state vector, $\mathbf{n}_t$, and a vector of observations, $\mathbf{y}_t$. As noted previously, point estimates of state vector components are typically quantities derived from sample data. We might instead use the raw sample data as observations. In contrast with the matrix model formulation, the general formulation presented here, which does not require a linear relationship between states and observations, can make modelling the raw observations more feasible.

The combination of the state model and the observation model is a state-space model and can be mathematically described as follows:

$$\text{Initial state pdf}: g_0(\mathbf{n}_0|\boldsymbol{\theta}) \tag{3.3}$$

$$\text{State } t \text{ pdf}: g_t(\mathbf{n}_t|\mathbf{n}_{t-1}, \boldsymbol{\theta}) \tag{3.4}$$

$$\text{Observation } t \text{ pdf}: f_t(\mathbf{y}_t|\mathbf{n}_t, \boldsymbol{\psi}), \tag{3.5}$$

where $\boldsymbol{\theta}$ is a vector of parameters corresponding to the state model, $\boldsymbol{\psi}$ is a vector of parameters corresponding to the observation model, and t=1,. . ., T. A sequence of state vectors, $\mathbf{n}_a$, $\mathbf{n}_{a+1}$, . . ., $\mathbf{n}_b$, will be denoted $\mathbf{n}_{a:b}$; $\mathbf{y}_{a:b}$ has a similar meaning.

The state process model, Eq. (3.4), is first-order Markov, i.e. the present state only depends on the previous state. Higher order Markov models can be re-written as first order Markov models by extending the dimension of the state vector to include previous states (see, for example, Schnute 1994). However, simply writing a univariate pdf for $\mathbf{n}_t$ as a function of previous states, e.g. $g_t(\mathbf{n}_t|\mathbf{n}_{t-1}, \mathbf{n}_{t-2}, \boldsymbol{\theta})$, does not pose any inference difficulties with modern approaches such as MCMC and sequential Monte Carlo, e.g., sequential importance sampling (see Chap. 4). The general class of such higher order Markov models was termed hidden process models by Newman et al. (2006).

Environmental, or temporal, variation in the state process can be made explicit by adding another level to the SSM for variability in survival, birth and other parameters.

$$\text{Parameter pdf} : h(\boldsymbol{\theta}_t|\boldsymbol{\Gamma}) \tag{3.6}$$

$$\text{Initial state pdf} : g_0(\mathbf{n}_0|\boldsymbol{\theta}_t) \tag{3.7}$$

$$\text{State } t \text{ pdf} : g_t(\mathbf{n}_t|\mathbf{n}_{t-1}, \boldsymbol{\theta}_t) \tag{3.8}$$

$$\text{Observation } t \text{ pdf} : f_t(\mathbf{y}_t|\mathbf{n}_t, \boldsymbol{\psi}). \tag{3.9}$$

Such a model is an example of a random effects or hierarchical state-space model (see Sect. 2.2.2). The pdf $h(\boldsymbol{\theta}_t|\boldsymbol{\Gamma})$ describes environmental stochasticity in $\boldsymbol{\theta}_t$, where $\boldsymbol{\Gamma}$ is a hyperparameter. Time-varying covariates could be used to model $\boldsymbol{\Gamma}$.

For Bayesian inference, another level is added to the state-space model formulation, namely, the prior pdf for the fixed and unknown parameters. For example, referring to the above hierarchical state-space model, Eqs. (3.6)–(3.9),

$$\text{Prior pdf} : \pi(\boldsymbol{\Gamma}, \boldsymbol{\psi}) \tag{3.10}$$

$$\text{Parameter pdf} : h(\boldsymbol{\theta}_t|\boldsymbol{\Gamma}) \tag{3.11}$$

$$\text{Initial state pdf} : g_0(\mathbf{n}_0|\boldsymbol{\theta}_t) \tag{3.12}$$

$$\text{State } t \text{ pdf} : g_t(\mathbf{n}_t|\mathbf{n}_{t-1}, \boldsymbol{\theta}_t) \tag{3.13}$$

$$\text{Observation } t \text{ pdf} : f_t(\mathbf{y}_t|\mathbf{n}_t, \boldsymbol{\psi}). \tag{3.14}$$

As will be described in more detail in Chap. 4, the end result of Bayesian inference for a state-space model is the joint posterior distribution for the parameters and the unknown states, i.e. $\pi(\mathbf{n}_{0:T}, \boldsymbol{\Gamma}, \boldsymbol{\psi}|\mathbf{y}_{1:T})$.

Finally, yet another layer of uncertainty is model uncertainty, also known as structural uncertainty (Williams et al. 2001). Alternative formulations for any of the above pdf's are often postulated. Competing theories about the science underlying the population dynamics translate into different state pdf's, g_t, or parameter pdf's, h. Denoting a particular model by $\mathcal{M}$, the Bayesian hierarchical model of Eqs. (3.10)–(3.14) can be extended as follows.

$$\text{Model prior pdf}: r(\mathscr{M}) \tag{3.15}$$

$$\text{Parameter prior pdf}: \pi(\boldsymbol{\Gamma}, \boldsymbol{\psi}|\mathscr{M}) \tag{3.16}$$

$$\text{Parameter pdf}: h(\boldsymbol{\theta}_t|\boldsymbol{\Gamma}, \mathscr{M}) \tag{3.17}$$

$$\text{Initial state pdf}: g_0(\mathbf{n}_0|\boldsymbol{\theta}_t, \mathscr{M}) \tag{3.18}$$

$$\text{State } t \text{ pdf}: g_t(\mathbf{n}_t|\mathbf{n}_{t-1}, \boldsymbol{\theta}_t, \mathscr{M}) \tag{3.19}$$

$$\text{Observation } t \text{ pdf}: f_t(\mathbf{y}_t|\mathbf{n}_t, \boldsymbol{\psi}, \mathscr{M}). \tag{3.20}$$

The pdf's and associated parameters in Eqs. (3.16)–(3.20) are conditional on the model $\mathscr{M}$, where parameters and the pdf's can vary between models. Chapter 5 will discuss model selection and model averaging.

3.2 Examples of State-Space Models

Here we give four examples of SSMs with the state pdf g_t and observation pdf f_t fully specified. The first is a nonlinear and non-normal coho salmon SSM with a scalar state and a scalar observation variable. The other examples are stochastic extensions of examples from Chap. 2.

3.2.1 Simplified Salmon Example

In this simplified SSM,[1] the state variable is a scalar N_t, the number of juvenile salmon alive in year t (at some point in time in that year) in a particular river. The dynamics of N_t are modelled as a Poisson distribution version of a Ricker stock-recruitment model (Quinn and Deriso 1999).

$$N_t|N_{t-1} \sim \text{Poisson}\left(\alpha N_{t-1} e^{-\beta N_{t-1}}\right), \quad \alpha > 0, \beta > 0. \tag{3.21}$$

Implicitly, the parameter α includes survival between time $t-1$ and t, the proportion that are female, fecundity (number of eggs produced), and survival between egg deposition and juvenile life stage. For the deterministic version of the model at least, to avoid chaotic behaviour, the value of α needs to be less than 2.69, and to avoid cycling, α must be less than 2. The parameter β is a measure of density dependence: as β increases, density dependence increases. The equilibrium value is $\ln(\alpha)/\beta$.

[1] The state process in this salmon SSM is a considerable oversimplification of the population dynamics for most, maybe all, species of salmon. Typically juvenile salmon production in a given year is the result of spawning from two or more age classes (different cohorts) and those age classes were juveniles two or more years previously, so that $N_{t-2}, N_{t-3}, \ldots$ contribute to N_t.

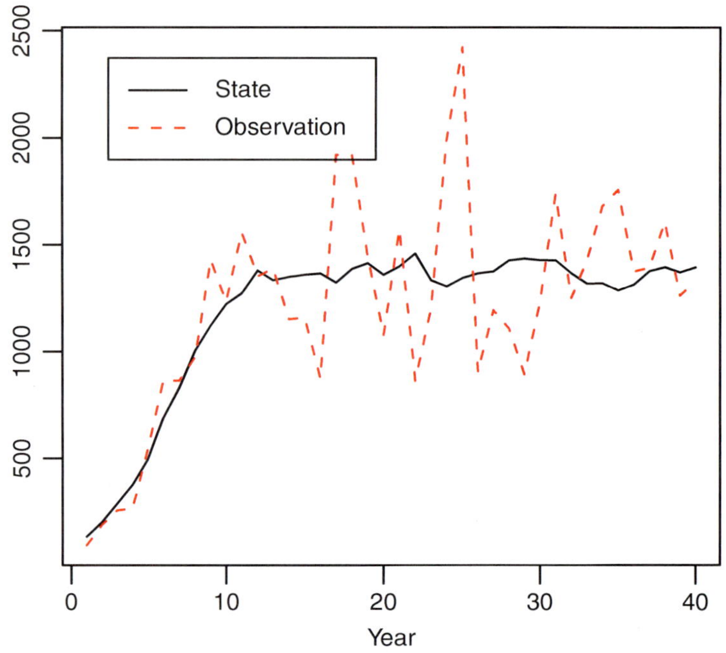

Fig. 3.2 Elementary salmon SSM with Ricker population dynamics with Poisson variation and lognormal observations. Ricker parameters are $\alpha = 1.5$ and $\beta = 0.0003$. Lognormal observations are bias-corrected with a coefficient of variation of 30 %

The observations are estimates of juvenile abundance, e.g. based upon samples take from in-river traps. A convenient probability distribution for non-negative valued observations is the lognormal distribution, although for counts of animals, continuous random variables are clearly an approximation:

$$y_t|N_t \sim \text{lognormal}\left(\log(N_t) - \sigma_y^2/2, \sigma_y^2\right). \tag{3.22}$$

This model is a bias-corrected lognormal, i.e. $E[y_t|N_t] = N_t$.

Forty years of simulated population dynamics and estimates are shown in Fig. 3.2. The estimates have a coefficient of variation of 30 %. (The code that generates this plot is given on the book website, see Sect. 1.2.)

3.2.2 BRS Model

For a more complex example, we return to the BRS formulation, an example of two states (e.g. immature and mature animals), summarized in Sect. 2.3. The sequence of sub-processes was survival (S), growth (R) and birth (B). Now we define stochastic processes for each of these sub-processes. The pdf g_t is difficult to evaluate analytically because it is a function of three different pdf's, but it is easy to describe symbolically and easy to simulate from. Symbolically, we write g_t as the following composite function (with the parameter vector $\boldsymbol{\theta}$ omitted to reduce notation):

$$g_t(\mathbf{n}_t|\mathbf{n}_{t-1}) = g_{3,t}\left(g_{2,t}\left(g_{1,t}(\mathbf{n}_{t-1})\right)\right). \tag{3.23}$$

The three pdf's, $g_{1,t}$, $g_{2,t}$, and $g_{3,t}$, represent the processes of survival, growth and birth. Statistically, this is a more useful formulation of the BRS model than is Eq. (2.11), because the latter tells us only the expected values of the states in $\mathbf{n}_t$, conditional on $\mathbf{n}_{t-1}$, whereas Eq. (3.23) represents the full joint distribution of the states in $\mathbf{n}_t$, conditional on $\mathbf{n}_{t-1}$.

The pdf $g_{1,t}(\mathbf{u}_{1,t}|\mathbf{n}_{t-1})$ corresponding to survival is the result of two binomial processes [Eq. (2.2)]:

$$\begin{pmatrix} u_{1(s),1,t} \sim \text{binomial}\left(n_{1,t-1}, \phi_1\right) \\ u_{1(s),2,t} \sim \text{binomial}\left(n_{2,t-1}, \phi_2\right) \end{pmatrix}. \tag{3.24}$$

Growth from immature to mature is another binomial process, so that $g_{2,t}(\mathbf{u}_{2,t}|\mathbf{u}_{1,t})$ is determined from Eq. (2.5):

$$\begin{pmatrix} u_{2(r),1,t} \sim \text{binomial}\left(u_{1(s),1,t}, 1-\pi\right) \\ u_{2(r),2,t} = u_{1(s),2,t} + \left(u_{1(s),1,t} - u_{2(r),1,t}\right) \end{pmatrix}. \tag{3.25}$$

If each adult can have at most one young, birth can be modelled as a third binomial process, so that $g_{3,t}(\mathbf{n}_t|\mathbf{u}_{2,t})$ is determined from a slight reformulation of Eq. (2.6):

$$\begin{pmatrix} n_{1,t} \equiv u_{3(b),1,t} = u_{2(r),1,t} + b_t \\ \quad \text{where } b_t \sim \text{binomial}(u_{2(r),2,t}, \rho) \\ n_{2,t} \equiv u_{3(b),2,t} = u_{2(r),2,t} \end{pmatrix}. \tag{3.26}$$

In contrast to evaluation of the state pdf $\mathbf{g}_t$, simulation from the distribution is easy: one simply simulates from g_1, that output is then input to simulate from g_2, and that output is then input to simulate from g_3. Suppose $\phi_1 = 0.50$, $\phi_2 = 0.71$, $\pi = 0.60$, and $\rho = 0.80$ and let the initial numbers, $\mathbf{n}_0$, be (50,70). A simulation of the population dynamics for the two size classes for 30 years is shown in Fig. 3.3. The eventual population growth in this example is exponential and density dependence is needed in survival or birth processes to stabilize the population. The observations $y_{I,t}$ (estimated number of immature animals) and $y_{M,t}$ (estimated number of mature animals) were taken to be lognormally distributed, unbiased, with a coefficient of variation of 30 %, i.e.

$$y_{I,t} \sim \text{lognormal}\left(\ln(n_{1,t}) - \sigma_y^2/2, \sigma_y^2\right)$$

$$y_{M,t} \sim \text{lognormal}\left(\ln(n_{2,t}) - \sigma_y^2/2, \sigma_y^2\right)$$

where $\sigma_y^2 = \ln(0.3^2 + 1)$.

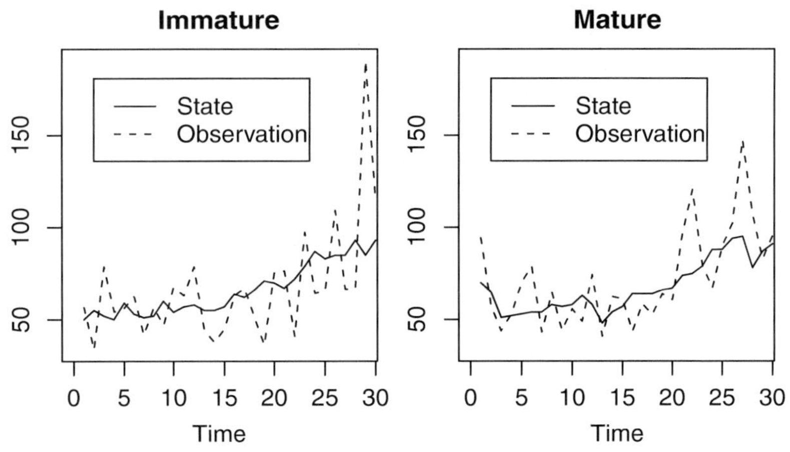

Fig. 3.3 Simulation of state process with two states, immature (I) and mature (M) animals, for $T = 30$ years, with three sub-processes, survival, growth and birth. Lognormally distributed estimates (the observations), with a coefficient of variation of 30 %, are also plotted

3.2.3 Coho Salmon

The processes survival, movement and harvest, that characterize the spatially partitioned coho salmon recoveries model described in Sect. 2.7.1, could be modelled using binomial (for survival and harvest) and multinomial (for movement) distributions. Here we describe a normal dynamic linear model (NDLM) approximation to those processes (Newman 1998):

$$\mathbf{n}_t | \mathbf{n}_{t-1} \sim \text{multivariate normal}\left(\mathbf{M}_t \mathbf{S}_{t-1} \mathbf{n}_{t-1}, \boldsymbol{\Sigma}_{n_t}\right), \quad t = 1, \ldots, 16 \tag{3.27}$$

$$\mathbf{y}_t | \mathbf{n}_t \sim \text{multivariate normal}\left(\mathbf{H}_t \mathbf{n}_t, \boldsymbol{\Sigma}_{y_t}\right), \quad t = 1, \ldots, 16. \tag{3.28}$$

The expected values are identical to the deterministic matrix models of Eqs. (2.18) and (2.22). The covariance matrix for the observations, $\boldsymbol{\Sigma}_{y,t}$, is a diagonal matrix with components $\sigma^2_{y,a,t}$, where

$$\sigma^2_{y,a,t} = n_{a,t} h_{a,t} (1 - h_{a,t}), \tag{3.29}$$

which is the variance for a binomial($n_{a,t}$, $h_{a,t}$) random variable. The components of the covariance matrix for the states, $\boldsymbol{\Sigma}_{n_t}$, can be constructed similarly using the variances and covariances of binomial random variables (for survival) and multinomial random variables (for movement); Newman (1998) gives a detailed example of the construction. Sullivan (1992) used a similar approach to constructing the covariance matrix in an NDLM approximation to binomial state processes.

3.2.4 Deer Metapopulation

The deer metapopulation model (Sect. 2.7.2) had five sub-processes characterizing the dynamics of the two deer population abundances. For convenience the expected state vector model is shown again below.

$$E[\mathbf{n}_t|\mathbf{n}_{t-1}] = \mathbf{CBAM}_t\mathbf{S}_{t-1}\mathbf{n}_{t-1}$$

Letting $\mathbf{n}'$ denote the transpose of the column vector $\mathbf{n}$,

$$\mathbf{n}'_t = [n_{A,f,1,t}, n_{A,m,1,t}, n_{A,f,2,t}, n_{A,f,3+,t}, n_{A,m,2+,t}, n_{B,f,1,t}, n_{B,m,1,t}, \\ n_{B,f,2,t}, n_{B,f,t}, n_{B,m,t}],$$

abundances distinguished by population (A or B), sex (f or m), and age class (1, 2 or 3+ for females, and 1 or 2+ for males). The matrices correspond to the processes of survival ($\mathbf{S}_{t-1}$), movement between populations A and B ($\mathbf{M}_t$), age incrementation ($\mathbf{A}$), births ($\mathbf{B}$), and assignment of sex ($\mathbf{C}$). Age incrementation is treated as a deterministic process, but the remaining processes can be modelled stochastically. Similar to the BRS model above, sub-process pdfs are specified, say g_S, g_M, g_A (a deterministic function), g_B and g_C. Consequently, the pdf $g_t(\mathbf{n}_t|\mathbf{n}_{t-1})$ is a composite function, analytically intractable, but potentially easy to simulate from using the individual component pdf's. Relatedly, and of relevance to issues of model fitting addressed in Chap. 4, analytic evaluation of probabilities of states is only tractable by separately evaluating the pdf's for intermediate or "latent" states, the "u's".

Here we describe one construction of $g_t(\mathbf{n}_t|\mathbf{n}_{t-1})$. Beginning with survival, survivors for each of the ten components of $\mathbf{n}_t$ are independent binomial random variables with various survival probabilities [e.g. Eq. (2.29)]. Given the cth component of $\mathbf{n}_{t-1}$, the survivors $u_{1(s),c,t}$ have the following distribution:

$$u_{1(s),c,t}|n_{c,t-1} \sim \text{binomial}\,(n_{c,t-1}, \phi_{c,t-1})\,. \tag{3.30}$$

The survivors in each component then move from their current population to the other population with probabilities specified previously; e.g. $\mu_{A\to B,t}$ specifies the probability that an animal moves from population A to population B, Eq. (2.30). The number moving is modelled as a binomial random variable; e.g. the movement of first year females from A to B,

$$u_{2(m),A\to B,f,1,t}|u_{1(s),A,f,1,t} \sim \text{binomial}\left(u_{1(s),A,f,1,t}, \mu_{A\to B,t}\right). \tag{3.31}$$

The new number in a given population (of sex s and age class a) is the sum of two independent binomial random variables, the "stayers" and the "movers", e.g. the first year females in B are those staying in B and those moving from A to B,

$$u_{2(m),B,f,1,t} = u_{2(m),A\to B,f,1,t} + u_{2(m),B\to B,f,1,t}\,. \tag{3.32}$$

As explained in Sect. 2.7.2, within each population, age incrementation shifts surviving first-year females to the second-year females group, surviving second-year and older females are combined (and labelled older females), and all surviving first-year and older males are combined (as older males). The fourth sub-process, birth, is modelled as a binomial process, with different success probabilities for second-year females and for older females, Eq. (2.32); e.g. the total births, males and females (total denoted ·) in population A, are the sum of births from the two sets of fecund females,

$$u_{4(b),A,\cdot,1,t} = u_{4(b),A,\cdot,1(2),t} + u_{4(b),A,\cdot,1(3+),t} \tag{3.33}$$

where

$$u_{4(b),A,\cdot,1(2),t} | u_{3(a),A,f,2,t} \sim \text{binomial}\left(u_{3(a),A,f,2,t}, \rho_1\right) \tag{3.34}$$

$$u_{4(b),A,\cdot,1(3+),t} | u_{3(a),A,f,3+,t} \sim \text{binomial}\left(u_{3(a),A,f,3+,t}, \rho_2\right). \tag{3.35}$$

The final sub-process, sex assignment, is treated as another binary (Bernoulli) process, where the number of females in the newly born animals is binomial; e.g. the number of first-year females in population A,

$$n_{A,f,1,t} | u_{4(b),A,\cdot,1,t} \sim \text{binomial}\left(u_{4(b),A,\cdot,1,t}, \alpha\right). \tag{3.36}$$

The observations were defined in Sect. 2.7.2 to be estimates of the total number of deer in each of the two populations, with no distinction between sex and age. Again lognormal distributions (with bias correction) are used; assuming independence between the population estimates,

$$y_{A,t} | \mathbf{n}_{A,\cdot,\cdot,t} \sim \text{lognormal}\left(\ln(\mathbf{n}_{A,\cdot,\cdot,t}) - \sigma_y^2/2, \sigma_y^2\right) \tag{3.37}$$

$$y_{B,t} | \mathbf{n}_{B,\cdot,\cdot,t} \sim \text{lognormal}\left(\ln(\mathbf{n}_{B,\cdot,\cdot,t}) - \sigma_y^2/2, \sigma_y^2\right) \tag{3.38}$$

where $\mathbf{n}_{A,\cdot,\cdot,t} = n_{A,f,1,t} + n_{A,m,1,t} + n_{A,f,2,t} + n_{A,f,3+,t} + n_{A,m,2+,t}$, and likewise for $\mathbf{n}_{B,\cdot,\cdot,t}$.

Chapter 4
Fitting State-Space Models

4.1 Introduction

This chapter is an overview of methods for using available data to make inferences about states and parameters of a state-space model. We call this "model fitting", or as Hilborn and Mangel (1997) say, "confronting models with data". Given a general SSM [Eqs. (3.3)–(3.5)],

$$\begin{aligned} \text{Initial state pdf} &: g_0(\mathbf{n}_0|\boldsymbol{\theta}) \\ \text{State } t \text{ pdf} &: g_t(\mathbf{n}_t|\mathbf{n}_{t-1}, \boldsymbol{\theta}) \\ \text{Observation } t \text{ pdf} &: f_t(\mathbf{y}_t|\mathbf{n}_t, \boldsymbol{\psi}), \end{aligned}$$

model fitting is a matter of using the data, $\mathbf{y}_{1:T}$, to estimate the unknown parameters, $(\boldsymbol{\theta}, \boldsymbol{\psi})$, or the unknown states, $\mathbf{n}_{0:T}$, or both—the dual estimation problem (Wan and Nelson 2001). At this point, we assume that the model structure has been specified; we later address model uncertainty. As noted previously, the definition of data, $\mathbf{y}_{1:T}$, is broad enough to include functions of sample data, e.g. point estimates of abundance and associated standard errors, as well as the raw data. The word "prediction", instead of estimation, is sometimes used when the unknown is a random variable instead of a constant; e.g. we predict the states $\mathbf{n}_{0:T}$ and estimate the parameters $(\boldsymbol{\theta}, \boldsymbol{\psi})$.

The literature on fitting state-space models is extensive, from the foundational work of Kalman (1960) and Kalman and Bucy (1961) to the present day, where research activity remains high. Means of fitting SSMs have advanced considerably in recent years, in particular using computer-intensive procedures. Taken together with advances in collecting data, we can now fit increasingly complex and realistic SSMs for population dynamics. This chapter aims to provide enough detail on these recent developments to allow the statistical ecologist to understand the pros and cons of the different fitting procedures. Additional details on specific SSM fitting algorithms and on available software are available in Commandeur et al. (2011).

K.B. Newman et al., *Modelling Population Dynamics: Model Formulation, Fitting and Assessment using State-Space Methods*, Methods in Statistical Ecology,
DOI 10.1007/978-1-4939-0977-3_4,

As a guide to readers with different interests, the organisation of the remainder of this chapter is as follows. Section 4.2 looks at the fitting, or more generally the inference, objectives from the perspective of an ecologist or natural resource manager, where we argue that attention needs to be focused on the state process model in ways similar to those presented for matrix models by Caswell (2001) and for population viability analysis by Morris and Doak (2002). Section 4.3 looks at fitting from a more purely statistical perspective (as given above), namely estimation of states and parameters conditional on the data. We discuss the central role of mathematical integration in inference for SSMs and highlight Bayesian and classical inference approaches to fitting SSMs. Section 4.4 describes "classical" statistical procedures for fitting SSMs, in particular the Kalman algorithm as it is applied to normal dynamic linear models, and the extended Kalman filter for nonlinear and non-normal SSMs. Section 4.5 discusses more recent stochastic algorithms for making inferences about SSMs, such as Markov chain Monte Carlo and Sequential Monte Carlo. Such stochastic procedures are more commonly used in a Bayesian framework but classical statistical inference can be carried out using these methods as well (e.g. de Valpine 2004).

4.2 An Ecological Perspective on Inference for SSMs

The primary SSM inferential objectives from an ecologist's or wildlife biologist's perspective are guided by the underlying scientific and management objectives. Different objectives can lead to emphases on estimating past states, predicting future states, estimating parameters related to vital rates such as survival, growth and reproduction, or quantifying the effects of environmental conditions or management actions on vital rates. Below we discuss a possible (and desirable) progression in inferential goals.

4.2.1 Knowing the Past and Present States

Often for a particular species or population, emphasis begins with the states, i.e. the initial questions about a population pertain to its abundance, its "status", and how those abundances have changed over time. As a Google search on the key word phrase "Population status and trends" reveals, status and trend reports for a large number of biological populations are ubiquitous products of natural resource management agencies. Before anything can be done, or should be done, about a population of interest, one needs to have some idea of the abundance of the population. That abundance might be further partitioned into different biological or geographical categories if the abundances of a structured population are of interest; e.g. number of age 1 females in different regions. For apparently healthy populations that are harvested for commercial, subsistence, or recreational purposes, estimates of current abundance, $\mathbf{n}_T$ say, are used to set harvest limits, quotas or regulations. For species that have been designated as threatened or endangered by some organization or agency such as the International Union for Conservation of Nature (IUCN) or the United States Fish and Wildlife Service, according to some criterion or law such as the U.S. Endangered Species Act, estimates of current and past abundances, $\mathbf{n}_{1:T}$, are used to determine if there are signs of population recovery, e.g. positive trends or positive step-changes in abundance.

4.2.2 Explaining the Past and Present, and Predicting the Future

Emphasis sometimes shifts, and we maintain that in many cases it should quickly do so, from estimating the states, or "status and trends", to estimating or quantitatively characterizing the underlying processes that drive the population dynamics. Even if one had perfect information about current and past states of the system, i.e. $\mathbf{n}_1, \ldots, \mathbf{n}_T$ were known without error, several questions would arise. Why did such a history occur? Why is the population trending downward toward extinction or quasi-extinction (Morris and Doak 2002)? Why did a positive step change occur? What will future states be? For natural resource managers, the most pressing question may be "What do I do now?", or "Given two alternate management actions, which one is better?". Predictions of future states and how different management actions affect those predictions can aid the decision-making process. The ideal model would take proposed actions A_j, $j = 1, \ldots, J$, and current system state, $\mathbf{n}_T$, as inputs and predict, without error, the next state $\mathbf{n}_{T+1}|A_j$. The manager could then compare $\mathbf{n}_{T+1}|A_j$, $j = 1, \ldots, J$, and select an action that was closest to the management objectives (with cost constraints also affecting the choice of action).

To understand the past and present and, similarly, to predict future system states, we need to know more than the system states: we need to understand the underlying dynamics that generated the states. Consider a simple univariate measure of the system, n_t, denoting the abundance of a single population. Given a sequence, $n_1, \ldots, n_T$, we can coarsely characterize the dynamics by $n_t = \lambda_{t-1} n_{t-1}$, where λ_{t-1} is the annual population growth rate. The sequence $\lambda_1, \ldots, \lambda_{T-1}$ can be quite informative in terms of characterizing the historic population growth rates. Estimates of such sequences are often the first step in carrying out a population viability analysis (PVA), "the use of quantitative methods to predict the likely future status of a population" (Morris and Doak 2002:1).

If we know the λ_ts, we would still like to know why they take those values. This requires knowing, or estimating, the underlying vital rates (survival probabilities and reproductive success, and possibly others) which are the basis of the λ_ts; e.g. what were the probabilities of annual survival for age 3 black bears for the last ten years? Thus estimation of underlying vital rates should be a next step; e.g. if $\lambda_t = (1 + \beta_t)\phi_t$, where β_t is the birth rate and ϕ_t is the survival probability, then estimation of β_t and ϕ_t is the next inferential objective.

Given estimates of vital rates, we might want to burrow down deeper to try to understand why the rates are what they are, to gain a more mechanistic understanding of the processes underlying the dynamics. In particular, we would like to understand the effects of both environmental covariates outside the direct control of humans, e.g. how winter temperatures affect bird survival (North and Morgan 1979), and deliberate management actions, e.g. crab fishery effects on Dungeness crab (Higgins et al. 1997). The quest for understanding the dynamics of a population is a continual one; the more information and understanding we gain, the more ideas will arise for managing or manipulating the population dynamics. For endangered populations in particular, there are always questions about what management actions yield the greatest improvement in population growth rates; e.g. increasing available habitat, improving the quality of current habitat, removing predators, increasing prey? We need not strive for the perfect model—if the mechanisms underlying vital rates can be explained to the degree that the effects of different management actions can be reliably predicted, this gives sufficient depth.

4.2.3 Remarks

- As SSMs become more widely used for modelling population dynamics, they will be used for similar purposes as are matrix population models (Chap. 2). Caswell (2001) and Morris and Doak (2002) demonstrate the use of matrix models to project population abundances into future years, to calculate probabilities of extinction (or quasi-extinction), to carry out sensitivity or elasticity type analyses for vital rates, and to assess the effects of environmental covariates or management actions on such vital rates and overall population growth rates.
- For ecologists and resource managers, the primary emphasis should be on the state process model, not on the observation model. While the type and quality of data are crucial to the complexity of an SSM that can be fitted, and while the observation model provides the essential link with the underlying state process model, data collection and the observation model simply provide imperfect snapshots of what the scientist and manager really need: the state vector. Even if the components of the state vector were known without error, the scientist still has substantial work to do to try to model the underlying dynamics. The science of population dynamics modelling is so challenging due to these dual problems: (1) collecting imperfect measurements on dynamic populations, e.g. using methods like mark-recapture or distance sampling or standard survey sampling procedures for fixed areas; and (2) trying to understand what causes the populations to vary as they do, e.g. whether there are density-dependent or environmental effects on survival probabilities or reproduction rates. It is important to keep a distinction in mind between the two problems, as we sometimes see field ecologists and biologists focused primarily on the data collection or population monitoring programmes and only secondarily on using those data to draw inferences about the underlying dynamics, and *vice versa* for theoretical ecologists.

- Simply postulating and mathematically formulating a state-space model for animal populations can be a useful exercise in its own right. That process alone can yield insights and greater clarity about the dynamics, the factors potentially affecting the dynamics, and which data are relevant. At the same time, it is relatively easy to formulate state process models in an SSM that are overly ambitious given the available data; e.g. a state process model for elk might project abundances by age and sex, but the only available data might be on total elk counts in September. In Chap. 5, we further discuss the merits of SSM formulation as well as the problem of data inadequacy, i.e. model over-specification. In this chapter, however, we assume that a "reasonable" SSM has been formulated in the sense that the parameters and states of the model are "estimable" given the available data, and we discuss different methods for making inferences about the SSM.

4.3 A Statistical Perspective on Inference for SSMs

As described at the beginning of this chapter, the general inferential objective is to characterize, or summarize, the unknown parameters of the SSM, $(\boldsymbol{\theta}, \boldsymbol{\psi})$, and the unknown states of the state process, $\mathbf{n}_{0:T}$, given the observations, $\mathbf{y}_{1:T}$. Sometimes the initial state, $\mathbf{n}_0$, is viewed as an unknown constant rather than the realisation of a stochastic process.

4.3.1 Inference as Integration

Inference for SSMs, whether for parameters, states, or both, ultimately involves integration, typically over a high-dimensional space (e.g. T integrals). To demonstrate this we begin with the *joint distribution of states and observations* of the most basic SSM in a non-Bayesian setting, namely Eqs. (3.3)–(3.5), which can be written as follows.

$$g_0(\mathbf{n}_0|\boldsymbol{\theta})g_{1:T}(\mathbf{n}_{1:T}, \mathbf{y}_{1:T}|\mathbf{n}_0, \boldsymbol{\theta}, \boldsymbol{\psi}) = g_0(\mathbf{n}_0|\boldsymbol{\theta}) \prod_{t=1}^{T} g_t(\mathbf{n}_t|\mathbf{n}_{t-1}, \boldsymbol{\theta}) f_t(\mathbf{y}_t|\mathbf{n}_t, \boldsymbol{\psi}). \quad (4.1)$$

Inference about parameters via maximum likelihood methods is based on the *marginal distribution of the observations*, equivalently the likelihood, thus integrating out the states in the joint distribution, i.e.

$$L(\boldsymbol{\theta}, \boldsymbol{\psi}|\mathbf{y}_{1:T}) \equiv f_{1:T}(\mathbf{y}_{1:T}|\boldsymbol{\theta}, \boldsymbol{\psi}) = \int g_0(\mathbf{n}_0|\boldsymbol{\theta}) \prod_{t=1}^{T} g_t(\mathbf{n}_t|\mathbf{n}_{t-1}, \boldsymbol{\theta}) f_t(\mathbf{y}_t|\mathbf{n}_t, \boldsymbol{\psi}) \, d\mathbf{n}_{0:T}. \quad (4.2)$$

Assuming the parameters are known constants, inferences about the states in the most general sense are based on the *conditional distribution of the states given the observations*, which also requires integration:

$$g_{0:T}(\mathbf{n}_{0:T}|\mathbf{y}_{1:T},\boldsymbol{\theta},\boldsymbol{\psi}) = \frac{g_0(\mathbf{n}_0|\boldsymbol{\theta})\prod_{t=1}^{T} g_t(\mathbf{n}_t|\mathbf{n}_{t-1},\boldsymbol{\theta}) f_t(\mathbf{y}_t|\mathbf{n}_t,\boldsymbol{\psi})}{f_{1:T}(\mathbf{y}_{1:T}|\boldsymbol{\theta},\boldsymbol{\psi})}. \tag{4.3}$$

We address handling of unknown parameters in a non-Bayesian framework in Sect. 4.3.2.4. A single summary measure of a particular state $\mathbf{n}_t$, such as the expected value, involves integration of the above conditional distribution over the remaining states, i.e.

$$E[\mathbf{n}_t|\mathbf{y}_{1:T},\boldsymbol{\theta},\boldsymbol{\psi}] = \int \mathbf{n}_t g_{0:T}(\mathbf{n}_{0:T}|\mathbf{y}_{1:T},\boldsymbol{\theta},\boldsymbol{\psi})\, d\mathbf{n}_{0:T}.$$

Bayesian inference is based on a larger joint distribution, extending Eq. (4.1) by adding a prior distribution for the parameters, which also requires integration; this is further explained in Sect. 4.3.2.

The ease with which inference can be carried out is then a function of how difficult it is to carry out the above integrations. In turn, the degree of difficulty of integration is a function of the particular formulation of the state and observation pdf's. The generality of the SSM structure invites creativity in the formulation of the pdf's, including the choice of probability distributions (e.g. normal, lognormal, Poisson), the nature of the parameterizations (e.g. linear, nonlinear), the dimensionality of the state and observation vectors, and the length of the time series, T.

Practical implementation of SSMs, however, only goes as far as the available integration techniques. Several integration techniques are currently available and have been applied to SSMs. These methods include exact analytical (e.g. Kalman filter), approximate analytical (e.g. extended Kalman filter), numerical, and Monte Carlo simulation methods. This listing is roughly a historical progression of techniques and SSM complexity approximately parallels this progression, i.e. SSMs for which inferences can be made using exact analytical methods are simpler than SSMs which can be fitted with Monte Carlo simulation methods. Statistical software developed for fitting SSMs of varying distributional structure and levels of complexity has progressed considerably over the past decade. An entire issue of the *Journal of Statistical Software* (May 2011, Volume 11) was devoted to the topic of fitting SSMs and the introductory article by Commandeur et al. (2011) is recommended reading. Here we discuss some of these methods in relatively broad terms in Sects. 4.4.1 (exact and approximate analytical methods, particularly the Kalman filter), 4.4.3 (numerical procedures), and 4.5 (Monte Carlo simulation procedures). Before doing so, we discuss the distinction between Bayesian and classical statistical approaches as applied to fitting SSMs.

4.3.2 *Bayesian Versus Classical Inference for SSMs*

This is a brief overview of the distinction between Bayesian and classical statistical procedures. We also look at how the approaches differ in the context of SSMs. Additional details can be found in many excellent references (e.g. Gelman et al. 2003; Carlin and Louis 2009; King et al. 2009).

4.3.2.1 Bayes vs Classical

Underlying both Bayesian and classical statistical inference is the likelihood, i.e., the joint probability distribution for the observed data viewed as a function of the unknown parameters. As a simple example of a likelihood, consider two passes of removal sampling of trout in a small stream (see Chap. 5 of Borchers et al. 2002). A net is swept on two consecutive occasions through the stream section, and each trout is caught with probability p; the captures are independent within and between sweeps, and trout that are captured are not returned to the stream. The observations are the numbers caught in the first and second pass, denoted y_1 and y_2. The parameters are p and N, the total number of fish present. The probability distribution for y_1 is binomial(N, p) and the distribution for y_2, conditional on the number y_1 removed previously, is binomial($N - y_1, p$). Thus the joint distribution is the product of the two binomial distributions:

$$f(y_1, y_2|N, p) \equiv f(\mathbf{y}|\boldsymbol{\theta}) = \binom{N}{y_1} p^{y_1}(1-p)^{N-y_1} \times \binom{N - y_1}{y_2} p^{y_2}(1-p)^{N-y_1-y_2}. \tag{4.4}$$

Classical statistical inference for the unknown parameters is usually centred on the likelihood. The expression in Eq. (4.4) is regarded as a function of the parameters given the observations: $f(\mathbf{y}|\boldsymbol{\theta}) \equiv L(\boldsymbol{\theta}|\mathbf{y})$, where $\boldsymbol{\theta} = (N, p)$. Conceptually, the parameters are considered to have fixed but unknown values. The most common point estimate of a parameter is the maximum likelihood estimate (mle); the likelihood $L(\boldsymbol{\theta}|\mathbf{y})$ has its maximum value when the mles of the parameters $\boldsymbol{\theta} = (N, p)$ are substituted in. Referring to the removal sampling example, given particular values of y_1 and y_2, the mle's for N and p are those values that maximize Eq. (4.4). These values are

$$\hat{N} = \frac{y_1^2}{y_1 - y_2}$$

$$\hat{p} = \frac{y_1 - y_2}{y_1}.$$

Note that $\hat{N}$ is only useful when $y_1 > y_2$; when $y_1 = y_2$, $\hat{N}$ is undefined, and when $y_1 < y_2$, the calculated values of $\hat{N}$ and $\hat{p}$ are negative.

Bayesian inference differs in a fundamental sense. Before collecting or analysing data, we specify probability distributions that quantify our beliefs about the values of unknown parameters. Namely, we specify a prior distribution, $\pi(\boldsymbol{\theta})$, where the parameters $\boldsymbol{\theta}$ are now considered to be random variables rather than having fixed but unknown values. Then given the data, we modify our beliefs to produce a posterior distribution for the parameters, $\pi(\boldsymbol{\theta}|\mathbf{y})$, where $\mathbf{y}$ represents the data. The posterior distribution is simply a conditional probability distribution, and Bayes formula is used to describe this distribution:

$$\begin{aligned}\pi(\boldsymbol{\theta}|\mathbf{y}) &= \frac{\pi(\boldsymbol{\theta},\mathbf{y})}{f(\mathbf{y})} = \frac{f(\mathbf{y}|\boldsymbol{\theta})\pi(\boldsymbol{\theta})}{\int f(\mathbf{y}|\boldsymbol{\theta})\pi(\boldsymbol{\theta})d\boldsymbol{\theta}} \\ &\propto f(\mathbf{y}|\boldsymbol{\theta})\pi(\boldsymbol{\theta}). \end{aligned} \tag{4.5}$$

Thus Bayesian inference is based on both the likelihood, $f(\mathbf{y}|\boldsymbol{\theta}) \equiv L(\boldsymbol{\theta}|\mathbf{y})$, and the prior distribution for the parameters, $\pi(\boldsymbol{\theta})$.

Returning to the removal sampling example, we first specify a prior distribution for the parameters N and p. For example,

$$\begin{aligned} N &\sim \text{discrete uniform}(a, b) \\ p &\sim \text{uniform}(0, 1) \end{aligned}$$

where a and b are positive integers such that $a < b$. The discrete uniform (a, b) distribution is such that for $x = a, a+1, \ldots, b$, $\Pr(N = x) = 1/(b - a + 1)$. The posterior distribution for N and p is then:

$$\begin{aligned}\pi(N, p|y_1, y_2) \propto{}& \frac{1}{b-a+1}\binom{N}{y_1}p^{y_1}(1-p)^{N-y_1} \\ &\times \binom{N-y_1}{y_2}p^{y_2}(1-p)^{N-y_1-y_2}\end{aligned}$$

for $N = a, a+1, \ldots, b$ and $0 \le p \le 1$.

Bayesian methods have some advantages over classical methods. One is that previous knowledge, independent of the current data, can be explicitly combined with the new data. One perspective on Bayes theorem is that, given the current data, we update or revise our prior knowledge of a process or phenomenon. Further by constraining the support (the set of allowable values) via a prior distribution, we ensure that the corresponding posterior distribution does not have support for nonsensical parameter values. As an example of the latter problem, for the removal sample example, maximum likelihood estimates are inadmissible when $y_1 \le y_2$. Adopting a Bayesian approach, we can select a prior distribution for N

that constrains it to lie in the interval (a, b), thus ensuring that posterior estimates (both point and interval) remain in the permissible range.

A further advantage of Bayesian methods is that the posterior distribution is necessarily a more complete description of information about a specific parameter than classical point estimates (e.g. maximum likelihood estimates) and confidence intervals.

A potential disadvantage of Bayesian methods, particularly in the case of relatively small samples, is that the influence of the prior distribution on the posterior distribution can be sizable. As is sometimes said, "the prior is swamping the data". Individuals with vastly different prior distributions might end up with vastly different posterior distributions and arrive at conclusions that differ in important ways. Selection of non-informative or neutral priors has been a topic of considerable discussion and research (Gelman et al. 2003).

4.3.2.2 Hierarchical Models

The terminology used to describe both hierarchical models and Bayesian and classical inference for such models can be confusing. We devote time to this issue because SSMs are a special case of hierarchical models and classical and Bayesian inference procedures for SSMs are discussed next. To try to clear up potential confusion, we consider the following simple model.

$$\alpha_i \sim \text{normal}\left(\psi, \tau^2\right), \; i = 1, \ldots, p \tag{4.6}$$

$$y_{i,j} | \alpha_i \sim \text{normal}\left(\alpha_i, \sigma^2\right), \; j = 1, \ldots, q_i. \tag{4.7}$$

To make the example more concrete, suppose there are p lakes which were sampled for a species of fish one summer, and in lake i, q_i fish were caught and measured for lengths. The observation, $y_{i,j}$, is the length of the jth fish caught in the ith lake. The average lengths are allowed to differ between lakes.

Such models have been given several different names, including hierarchical, multi-level, random effects, and variance components models. The equivalence between hierarchical and multi-level is likely apparent. The model for the observations, $y_{i,j}$, can be viewed as a lower level model, while the model for the lake-specific means, α_i, is a higher level model. The random effects label refers to the random variable α_i which then becomes a parameter in the model for the observations, $y_{i,j}$. The label variance components refers to the terms τ^2 and σ^2.

Inference objectives include estimation of the fixed parameters, ψ, τ^2, and σ^2, and the random variables, $\alpha_i, i{=}1, \ldots, n$. A common classical inference approach to estimating the fixed parameters is maximum likelihood. The likelihood is evaluated by integrating over the unobserved α_i,

$$L(\psi, \tau^2, \sigma^2) = \prod_{i=1}^{p} \int_{\alpha_i} \left[\prod_{j=1}^{q_i} \Phi\left(\frac{y_{i,j} - \alpha_i}{\sigma}\right) \right] \Phi\left(\frac{\alpha_i - \psi}{\tau}\right) d\alpha_i$$

where $\Phi()$ denotes the standard normal density function. We note that maximum likelihood estimates of the variance components, τ^2 and σ^2, are typically biased, and REML (restricted maximum likelihood) (Harville 1977) is an alternative procedure which yields less biased or unbiased estimates.

Classical inference estimates of the random variable, α_i, can be made conditional on the estimated parameters and the observations, namely,

$$\hat{\alpha}_i = E\left[\alpha_i|\mathbf{y}, \hat{\psi}, \hat{\tau}^2, \hat{\sigma}^2\right] = \left[\frac{\hat{\sigma}^2/q_i}{\hat{\sigma}^2/q_i + \hat{\tau}^2}\right]\hat{\psi} + \left[\frac{\hat{\tau}^2}{\hat{\sigma}^2/q_i + \hat{\tau}^2}\right]\overline{y}_i. \quad (4.8)$$

Such estimates are called Empirical Bayes estimates (Casella 1985). This terminology, *Empirical Bayes*, may seem somewhat confusing in that a prior distribution was not explicitly defined in the original hierarchical model formulation. However, an alternative perspective on the formulation is that the model of Eq. (4.6) is a Bayesian prior distribution for the α_i (Casella 1985), where the values ψ and τ^2 were fixed in advance, arbitrarily. Purely Bayesian point estimates for α_i are the means of the posterior distribution, which equal Eq. (4.8) with the fixed values substituted (assuming for simplicity here that σ^2 was known).

We note that an alternative Bayesian inference procedure is to not assume known values for ψ and τ^2 and to specify a prior distribution, say $\pi(\psi, \tau^2)$ (again assume σ^2 is known, though a prior distribution could be chosen for it). Then the posterior distribution for ψ and τ^2 would be calculated,

$$\pi(\psi, \tau^2|\mathbf{y}) \propto \pi(\psi, \tau^2)\prod_{i=1}^{p}\int_{\alpha_i}\left[\prod_{j=1}^{q_i}\Phi\left(\frac{y_{i,j} - \alpha_i}{\sigma}\right)\right]\Phi\left(\frac{\alpha_i - \psi}{\tau}\right)d\alpha_i.$$

All information about the α_i would then be found in the conditional distribution for α_i given $\mathbf{y}$ and the prior distribution $\pi(\psi, \tau^2)$. Such inferences are called, somewhat confusingly again, Bayes empirical Bayes (Carlin and Louis 2009).

Finally, Clark (2005) makes a distinction between the meaning of randomness as it applies to the α_i and to ψ and τ^2. The α_i are inherently random variables, the values are varying in some temporal or spatial sense, e.g., the average lengths are varying between lakes. The probability distribution for α_i quantifies that inherent randomness. Conversely ψ and τ^2 are fixed constants but their values are unknown. A prior distribution for these fixed constants is a reflection of *uncertainty*, ignorance, say, not a reflection of temporal or spatial variability in their values. Bayesian statisticians may or may not make such a distinction and simply refer to the α_i and ψ and τ^2 as random variables.

4.3.2.3 Bayesian Inference for SSMs

In the case of state-space models and the dual estimation objective, Bayesian inference involves calculating, or generating a sample from, the posterior distribution for the parameters and the states. Prior distributions are needed for the parameters, say $\pi(\boldsymbol{\theta}, \boldsymbol{\psi})$. Unless the initial state value, $\mathbf{n}_0$, is assumed known, then a prior distribution (the initial state distribution) must also be specified. Priors for the unknown, random states, $\mathbf{n}_{1:T}$, need not be specified because the state process model itself implicitly defines their prior distribution. The Bayes formula for a state-space model can be written as follows.

$$\begin{aligned}\pi(\mathbf{n}_{0:T}, \boldsymbol{\theta}, \boldsymbol{\psi}|\mathbf{y}_{1:T}) &\propto \pi(\mathbf{n}_{0:T}, \boldsymbol{\theta}, \boldsymbol{\psi}, \mathbf{y}_{1:T}) \\ &= \pi(\boldsymbol{\theta}, \boldsymbol{\psi}) g_0(\mathbf{n}_0|\boldsymbol{\theta}) \prod_{t=1}^{T} g_t(\mathbf{n}_t|\mathbf{n}_{t-1}, \boldsymbol{\theta}) f_t(\mathbf{y}_t|\mathbf{n}_t, \boldsymbol{\psi}), \quad (4.9)\end{aligned}$$

where the SSM is assumed first-order Markov.

Analytic calculation of the posterior distribution $\pi(\mathbf{n}_{0:T}, \boldsymbol{\theta}, \boldsymbol{\psi}|\mathbf{y}_{1:T})$ is prohibitively difficult, usually impossible, for most SSMs. The difficulty lies in the denominator of Bayes formula [Eq. (4.5)], $f(\mathbf{y})$, which is typically an intractable integral. Numerical integration methods can be used for small (say two or three) dimensional problems, but most of these methods fail in higher dimensions.

Bayesian inference has been made feasible for high-dimensional models, including SSMs, through the combination of Monte Carlo or simulation-based algorithms and high-speed computing. These computer-intensive Monte Carlo procedures can be used to produce samples from the posterior distribution. The best known and most widely used Monte Carlo sampling procedure is Markov chain Monte Carlo (MCMC). `WinBUGS` is a freely available software implementation of MCMC and is the tool we have chosen for demonstrating Bayesian inferences in this book. The open-source version of `WinBUGS` is `OpenBUGS`, and another free option is the `JAGS` software. Alternative Monte Carlo procedures designed primarily for dynamic stochastic models, including SSMs, are Sequential Monte Carlo methods with a specific approach being Sequential Importance Sampling (SIS).

4.3.2.4 Classical Inference for SSMs

In Sect. 4.4, we provide specifics on various classical statistical inference procedures for SSMs, in particular the highly influential Kalman filter algorithm. Here discussion is limited to brief general principles.

The classical approach differs from the Bayesian approach in that there are no prior distributions for the unknown parameters. Consider the following simple SSM where n_t and y_t are scalars, and where parameters from the observation model are removed to reduce notation,

$$g_t(n_t|n_{t-1}, \boldsymbol{\theta})$$

$$f_t(y_t|n_t).$$

Inference often proceeds in two stages: first, the unknown fixed parameters ($\boldsymbol{\theta}$) are estimated, then estimates (sometimes called predictions) of the unknown states are made conditional on the fixed parameter estimates (implicitly, on the observations). Assuming that estimation of the parameters is via maximum likelihood, calculation of the marginal distribution for the observations is required, i.e. integration over the unknown states is carried out [Eq. (4.2)]. Treating these parameter estimates as constants (i.e. ignoring parameter uncertainty), the conditional distribution for the states [Eq. (4.3)] is a complete summary of information about the unknown states.

$$g_{0:T}(\mathbf{n}_{0:T}|\mathbf{y}_{1:T}, \widehat{\boldsymbol{\theta}}) = \frac{g_0(\mathbf{n}_0|\widehat{\boldsymbol{\theta}}) \prod_{t=1}^{T} g_t(\mathbf{n}_t|\mathbf{n}_{t-1}, \widehat{\boldsymbol{\theta}})}{f_{1:T}(\mathbf{y}_{1:T}|\widehat{\boldsymbol{\theta}})}. \tag{4.10}$$

Estimated expected values for individual states can be used as point estimates and interval estimates can be based on percentiles of Eq. (4.10).

We note that this two-stage procedure is a special case of what is sometimes called parametric empirical Bayes (PEB) methods (Morris 1983). The label "Bayes" is potentially confusing as the procedure is not Bayesian. PEB methods are commonly used for random effects models. For example, consider the following simple random effects model where we retain some of the SSM notation (removing additional parameters from the observation model for simplicity):

$$\mathbf{n} \sim g(\mathbf{n}|\boldsymbol{\theta})$$
$$\mathbf{y} \sim f(\mathbf{y}|\mathbf{n}).$$

Both $\mathbf{n}$ and $\mathbf{y}$ are random vectors (or scalars), where $\mathbf{y}$ are observed and $\mathbf{n}$ are unknown and the random effects. Again the parameters are estimated first. Second, the conditional distribution for the random effects given the observations is calculated by substituting the mle's, say $\hat{\boldsymbol{\theta}}$, for the parameters:

$$g\left(\mathbf{n}|\mathbf{y}, \hat{\boldsymbol{\theta}}\right).$$

This conditional distribution is also known as the estimated posterior distribution (Carlin and Louis 1996).

One key detail glossed over here is the degree of difficulty in carrying out the integrations, to yield the likelihood, the estimated posterior distribution, and parametric Empirical Bayes estimates. Skaug (2002), for example, uses the Laplace approximation to carry out some of the integration. A second key detail is the initial state, $\mathbf{n}_0$, whether it is viewed as a random effect (as shown here) or as a parameter (a constant), and in either case whether or not it is in fact estimable. Some discussion of how to handle $\mathbf{n}_0$ in the special case of normal dynamic linear SSMs is given in Sect. 9.5.1. Here we note that, given a long enough time series, the initial observation $\mathbf{y}_1$ can be used to estimate $\mathbf{n}_1$ in a somewhat *ad hoc* manner by inverting the observation model and solving for $\mathbf{n}_1$, and no inferences about $\mathbf{n}_0$ are made. A simple example of such an inversion would be an observation model where a scalar y_1 is proportional to n_1 (also scalar), say $y_1 \sim \text{normal}(\phi n_1, \sigma_y^2)$ for some unknown ϕ. Given an initial value for ϕ, n_1 is estimated by y_1/ϕ (Thomas et al. 2005, used this approach in a Bayesian setting to formulate a prior for the initial state value).

4.4 Classical Statistical Procedures for Fitting SSMs

The original discrete time SSM formulated by Kalman (1960) uses normal distributions of states and observations and the conditional expected values for states and for observations are linear combinations of previous states and current states, respectively, i.e. a normal dynamic linear model (NDLM). He developed a closed-form analytic algorithm, the Kalman filter, for doing the integration to calculate both the likelihood and the conditional distributions of the states, which has led to extremely widespread use for a broad range of problems. Additional explanations of the Kalman algorithms can be found in Meinhold and Singpurwalla (1983) and Harvey (1989).

4.4.1 *The Kalman Filter*

The Kalman filter is designed for fitting NDLMs. Even though stochastic integration techniques are more flexible because they allow fitting nonlinear, non-normal models, an overview of Kalman's algorithms is important on several counts, including their continued popular usage, relative simplicity and speed.

In a NDLM, the state and observation processes are normal random variables and the expected values are linear combinations of the conditioning variable, either $\mathbf{n}_{t-1}$ or $\mathbf{n}_t$. More concisely,

$$\mathbf{n}_0 \sim \text{normal}(\boldsymbol{\mu}_0, \mathbf{Q}_0) \tag{4.11}$$

$$\mathbf{n}_t|\mathbf{n}_{t-1} \sim \text{normal}(\mathbf{A}_{t-1}\mathbf{n}_{t-1}, \mathbf{Q}_t) \tag{4.12}$$

$$\mathbf{y}_t|\mathbf{n}_t \sim \text{normal}(\mathbf{B}_t\mathbf{n}_t, \mathbf{R}_t), \tag{4.13}$$

where $\mathbf{A}_{t-1}$ is an $m \times m$ matrix of constants, $\mathbf{B}_t$ is a $k \times m$ matrix of constants with k the number of observations in year t, and $\mathbf{Q}_t$ and $\mathbf{R}_t$ are covariance matrices for $\mathbf{n}_t$ and $\mathbf{y}_t$, respectively. Often the initial state vector, $\mathbf{n}_0$, is assumed to be a fixed value.

NDLMs may be viewed as approximations to more realistic nonlinear, non-normal SSMs for the dynamics underlying animal populations and the type of data collected from such populations. The existence of the Kalman filter (KF) makes NDLMs attractive. The KF is a recursive analytic procedure for estimating the states $\mathbf{n}_t$, given the observations up to and including time t, $\mathbf{y}_{1:t}$. In particular, the KF yields

T probability distributions for the unobserved states, n_t, $t = 1, \ldots, T$, conditional on the observations $y_{1:t}$, i.e. $f(n_t|y_{1:t})$, the so-called filtered distribution. A related algorithm, the Kalman smoother, calculates T probability distributions for the states conditional on *all* the observations, i.e. $f(n_t|y_{1:T})$. The resulting distributions in both cases are normal, thus the algorithms only need to calculate the mean vector and variance-covariance matrices. The KF can also be used to calculate the likelihood (the marginal distribution for the observations), which is necessary for calculation of maximum likelihood estimates. In comparison to computer-intensive procedures like MCMC and SIS, the Kalman filter can be very fast.

The Kalman filter is explained assuming that parameters of the pdf are known, i.e. the matrices $\mathbf{A}_t, \mathbf{B}_t, \mathbf{Q}_t, \mathbf{R}_t$ in Eqs. (4.11)–(4.13) are known. A web page devoted to Kalman algorithms which includes links to expository articles is http://www.cs.unc.edu/~welch/kalman/. The filter is a recursive algorithm in that values calculated for time $t-1$ are used to calculate values for time t. At each step of the recursion, there are two sub-steps: (a) predicting $\mathbf{n}_t$ using an estimate of $\mathbf{n}_{t-1}$; (b) "updating" $\mathbf{n}_t$ using a weighted combination of the prediction and the observation for time t, $\mathbf{y}_t$. Both the predicting and the updating steps have corresponding calculations for the variance-covariance matrices of the predictions and updated estimates.

Notation for the predicted state and its covariance matrix is $\mathbf{n}_t^{t-1}$ and $\mathbf{P}_t^{t-1}$, while that for the updated (or "filtered") state and its covariance matrix is $\mathbf{n}_t^t$ and $\mathbf{P}_t^t$.

Begin at $t = 0$ with fixed values for $\mathbf{n}_0$ and $\mathbf{P}_0^0$; often $\mathbf{P}_0^0$ is set equal to zero. Then predict the state at $t = 1$ using the state pdf, Eq. (4.12), and calculate the corresponding covariance matrix:

$$\text{Prediction of } \mathbf{n}_t\text{:} \quad \mathbf{n}_t^{t-1} = \mathbf{A}_{t-1}\mathbf{n}_{t-1}^{t-1} \tag{4.14}$$

$$\text{Covariance matrix of } \mathbf{n}_t^{t-1}\text{:} \quad \mathbf{P}_t^{t-1} = \mathbf{A}_{t-1}\mathbf{P}_{t-1}^{t-1}\mathbf{A}_{t-1}' + \mathbf{Q}_t. \tag{4.15}$$

Next update the state and calculate the corresponding covariance matrix:

$$\text{Update of } \mathbf{n}_t\text{:} \quad \mathbf{n}_t^t = \mathbf{n}_t^{t-1} + \mathbf{K}_t(\mathbf{y}_t - \mathbf{B}_t\mathbf{n}_t^{t-1}) \tag{4.16}$$

$$\text{Covariance matrix of } \mathbf{n}_t^t\text{:} \quad \mathbf{P}_t^t = (\mathbf{I} - \mathbf{K}_t\mathbf{B}_t)\mathbf{P}_t^{t-1}, \tag{4.17}$$

where $\mathbf{K}_t$ is referred to as the Kalman gain and is defined by

$$\mathbf{K}_t = \mathbf{P}_t^{t-1}\mathbf{B}_t'(\mathbf{B}_t\mathbf{P}_t^{t-1}\mathbf{B}_t' + \mathbf{R}_t)^{-1}.$$

Increment t by 1, repeat the calculations in Eqs. (4.14)–(4.17), and stop after finishing the updating for time T.

4.4.1.1 Numerical Demonstration of the Kalman Filter

The following simple NDLM is used to demonstrate the KF. The state vector has just two components. The matrix $\mathbf{A}_t$ is constant over time, as is the covariance matrix $\mathbf{Q}_t$.

Table 4.1 Observations (y_t), true unobserved states (n_t), and filtered states (n_t^t) from a simulated NDLM with $T = 10$ observations

Juveniles											
t	0	1	2	3	4	5	6	7	8	9	10
y_t	0	62	45	51	48	53	50	52	53	49	49
n_t	30	75	57	66	62	64	64	64	65	64	64
n_t^t	30	75	55	64	60	63	62	63	63	62	62
Adults											
t	0	1	2	3	4	5	6	7	8	9	10
y_t	0	42	48	46	45	44	45	46	44	45	47
n_t	50	38	44	41	43	43	43	43	42	43	43
n_t^t	50	37	43	40	41	41	41	41	42	42	42

Suppose that the $n_{t,1}$ are juveniles and $n_{t,2}$ are adults and the matrix $\mathbf{A}_t$ is a Leslie matrix. Let the state process equation be

$$\begin{bmatrix} n_{t,1} \\ n_{t,2} \end{bmatrix} = \begin{bmatrix} 0.0 & 1.5 \\ 0.3 & 0.55 \end{bmatrix} \begin{bmatrix} n_{t-1,1} \\ n_{t-1,2} \end{bmatrix} + \begin{bmatrix} q_{t,1} \\ q_{t,2} \end{bmatrix},$$

where $q_{t,1}$ and $q_{t,2}$ are assumed to be identically and independently distributed (iid) normal$(0, 0.5^2)$. We note that for this example we chose the values of the Leslie matrix carefully so that the deterministic rate of growth is exactly 1, i.e. deterministic projections of the population neither exponentially decline to extinction nor grow exponentially. This does not affect the main points we wish to demonstrate here, however.

Suppose that the observation vector has the same dimension as $\mathbf{n}_t$ and contains biased estimates of each component of $\mathbf{n}_t$, where the bias is assumed known.

$$\begin{bmatrix} y_{t,1} \\ y_{t,2} \end{bmatrix} = \begin{bmatrix} 0.8 & 0.0 \\ 0.0 & 1.1 \end{bmatrix} \begin{bmatrix} n_{t,1} \\ n_{t,2} \end{bmatrix} + \begin{bmatrix} r_{t,1} \\ r_{t,2} \end{bmatrix},$$

where $r_{t,1}$ and $r_{t,2}$ are iid normal$(0, 2^2)$.

Using R, the NDLM was simulated setting $\mathbf{n}_0 = (30, 50)$ and $\mathbf{P}_0^0 = 0$. The R code for simulation and using the Kalman filter is given on the book website (Sect. 1.2). Table 4.1 shows the simulated observations and states and the Kalman "filtered" estimates of the states. As is the case with Leslie matrix projection models (Caswell 2001), the abundances of juveniles and adults have, with the exception of relatively minor process variation, reached equilibrium values after four or five generations.

More commonly, the parameters are not known. The predicted state, $\mathbf{n}_t^{t-1}$, and corresponding covariance matrix, $\mathbf{P}_t^{t-1}$, can be used to construct the likelihood function (the marginal distribution for the observations). The likelihood can be written as follows (with conditioning on the initial state $\mathbf{n}_0$ made explicit).

$$L(\boldsymbol{\theta}, \boldsymbol{\psi} | \mathbf{y}_{1:T}, \mathbf{n}_0) = f(\mathbf{y}_1 | \mathbf{n}_0) \prod_{t=2}^{T} f(\mathbf{y}_t | \mathbf{y}_{1:t-1}, \mathbf{n}_0). \tag{4.18}$$

It can be shown that the conditional density for each $\mathbf{y_t}$ is normal:

$$\mathbf{y}_t | \mathbf{y}_{1:t-1}, \mathbf{n}_0 \sim \text{multivariate normal}\left(\mathbf{B}_t \mathbf{n}_t^{t-1}, \mathbf{B}_t \mathbf{P}_t^{t-1} \mathbf{B}_t' + \mathbf{R}_t\right). \tag{4.19}$$

Note that the mean is the expected value of $\mathbf{y}_t$ based on the observation equation, with the predicted value of $\mathbf{n}_t$ substituted for its true, but unknown value. Example R code is given on the book website (Sect. 1.2) for calculating the likelihood for the values in the matrix $\mathbf{A}$ alone and for calculating the maximum likelihood estimates of the parameters. Shumway and Stoffer (1982) show how to use the EM algorithm with the KF for simultaneous inference of states and parameters.

4.4.2 Extensions to NDLMs and the Kalman Filter

By 1961 the Kalman filter had been extended to the case of nonlinear, but normal, SSMs (Grewal and Andrews 2010):

$$\mathbf{n}_0 \sim \text{normal}(\boldsymbol{\mu}_0, \mathbf{Q}_0) \tag{4.20}$$

$$\mathbf{n}_t | \mathbf{n}_{t-1} \sim \text{normal}\left(\mathbf{g}(\mathbf{n}_{t-1}), \mathbf{Q}_t\right) \tag{4.21}$$

$$\mathbf{y}_t | \mathbf{n}_t \sim \text{normal}\left(\mathbf{f}(\mathbf{n}_t), \mathbf{R}_t\right), \tag{4.22}$$

where g and f are vectors of differentiable functions. The Extended Kalman Filter (EKF) substitutes linear (first order) Taylor series approximations for nonlinear state and/or observation equations in prediction and update steps (Eqs. (4.14)–(4.17)). For the prediction step, prediction of the $\mathbf{n}_t$ simply involves plugging the current updated value, $\mathbf{n}_{t-1}^{t-1}$, into g, while calculation of the corresponding covariance matrix, P_t^{t-1}, includes linearization of g:

$$\mathbf{n}_t^{t-1} = \mathbf{g}\left(\mathbf{n}_{t-1}^{t-1}\right)$$

$$\mathbf{P}_t^{t-1} = \mathbf{G}_{t-1} \mathbf{P}_{t-1}^{t-1} \mathbf{G}_{t-1}' + \mathbf{Q}_t,$$

where

$$\mathbf{G}_{t-1} = \left. \frac{\partial \mathbf{g}}{\partial \mathbf{n}} \right|_{\mathbf{n}_{t-1}^{t-1}}.$$

For the update step, the updated estimate is calculated as for the regular Kalman filter, but the updated estimated covariance involves a linearization of f:

$$\mathbf{n}_t^t = \mathbf{n}_t^{t-1} + \mathbf{K}_t (\mathbf{y}_t - \mathbf{B}_t \mathbf{n}_t^{t-1})$$

$$\mathbf{P}_t^t = (\mathbf{I} - \mathbf{K}_t \mathbf{F}_t) \mathbf{P}_t^{t-1},$$

where

$$\mathbf{F}_t = \left.\frac{\partial \mathbf{f}}{\partial \mathbf{n}}\right|_{\mathbf{n}_{t-1}^t}.$$

The realizable minimum variance (RMV) filter (due to Liang and Christensen 1978, but see the concise description in Quinn and Deriso 1999:236–237) is another procedure for handling nonlinear but normal SSMs. Julier et al. (1995) proposed the Unscented Kalman Filter (UKF) as a more accurate alternative to the EKF (see the review article by Wan and van der Merwe 2001).

The EKF and RMV filters also allow calculation of the likelihood of the parameters. Wan and van der Merwe (2001) describe two dual estimation procedures using the UKF, the dual UKF and the joint extended UKF.

A variation on NDLMs, conditionally Gaussian models, which can be analysed using the standard Kalman filter, are discussed in Sect. 9.7.1.

4.4.3 Numerical Procedures

When the SSM is nonlinear and/or non-normal, numerical procedures are sometimes used. We begin with a method developed by Kitagawa (1987). Kitagawa noted that prior to his work, the typical approach to handling nonlinear and non-normal SSMs was "to approximate the non-normal distribution by one or several normal distributions or by some parametric function"; e.g. the extended Kalman filter, the Gaussian (i.e. normal) sum filter (Alspach and Sorenson 1972). Kitagawa, however, worked directly with non-normal distributions of the following form:

$$\begin{aligned} n_t &= F n_{t-1} + G v_t \\ y_t &= H n_t + w_t, \end{aligned}$$

where for his numerical solution n_t and y_t were *scalars*, and v_t and w_t were non-normal random variables. He formulated iterative algorithms for evaluating one-step-ahead predictive density of the state (i.e. $g(n_t|\mathbf{y}_{1:t-1})$), filtered state density (i.e. $g(n_t|\mathbf{y}_{1:t})$), and "fixed" interval smoothed state density (i.e. $g(n_t, n_{t+1}|\mathbf{y}_{1:T})$). Given those algorithms, he numerically carried out evaluations of various conditional densities and convolutions of densities using piecewise linear (first-order spline) functions. Extensions to higher dimension state and observation vectors were discussed but not demonstrated.

A computationally efficient method for calculating maximum likelihood estimates of SSM parameters and then making inferences about unobserved states in high-dimensional nonlinear and/or non-normal SSMs, or more generally hierarchical models, has been developed by Skaug and Fournier (2006; see also Skaug 2002). They have developed freely-distributed companion software, ADMB-RE

(Automatic Differentiation Model Builder—Random Effects, Skaug and Fournier 2011), for fitting SSMs. Their approach has several features.

1. The SSM is viewed as a special case of a random effects model, i.e. a *hierarchical* model, where the unobserved states are the random effects.
2. The marginal likelihood for the observations [Eq. (4.2)] is calculated approximately (typically with high accuracy) and quickly using automatic differentiation (Skaug and Fournier 2006) and the Laplace method (Tierney and Kadane 1986) to carry out the high-dimensional integration over the state vector.
3. Given the marginal likelihood, maximum likelihood estimates of the fixed SSM parameters are calculated (although Bayesian inference is an option).
4. Parametric empirical Bayes (Morris 1983) estimates of the random effects, i.e. the unobserved states, are calculated conditional on the maximum likelihood estimates, i.e. given the mle's for the parameters, ADMB-RE "automatically calculates 'maximum posterior' estimates of the random effects" (Skaug and Fournier 2011).

4.5 Monte Carlo Simulation Procedures

Computer-intensive Monte Carlo procedures such as Markov chain Monte Carlo (MCMC) and sequential importance sampling (SIS) are more commonly used within a Bayesian framework, but they can also be used for classical analyses (Geyer 1996; de Valpine 2002, 2003, 2004; de Valpine and Hilborn 2005; Ionides et al. 2006). Below we discuss MCMC and SIS as distinct approaches but note that the methods can be used in combination, as in Partical MCMC (Andrieu et al. 2010).

4.5.1 Markov Chain Monte Carlo

The general idea of Markov chain Monte Carlo (MCMC) is to generate samples from a Markov chain which has a limiting distribution equal to the desired distribution (Metropolis et al. 1953; Hastings 1970; Gilks et al. 1996; Brooks et al. 2011). The iterative procedure known as the Metropolis–Hastings algorithm is a means of constructing the appropriate Markov chain. Before describing the algorithm for the case of an SSM with unknown parameters, we begin with a simpler case of a sample of n independent identically distributed random variables, $y_1, \ldots, y_n$ with pdf $f(\mathbf{y}|\theta)$, where θ is an unknown scalar and $\mathbf{y} = \{y_1, \ldots, y_n\}$. Let $\pi(\theta)$ be the prior for an unknown parameter and let $f(\mathbf{y}|\theta)$ denote the likelihood. The Metropolis–Hastings procedure generates a sample from the posterior distribution for θ, i.e. a sample from $\pi(\theta|\mathbf{y})$, in the following manner. At each iteration i of the algorithm, let the current state of the chain be denoted by θ^i. A *candidate* value, denoted θ', is generated from some *proposal* distribution with pdf $q(\theta'|\theta^i)$. The proposed parameter is accepted with probability $\alpha(\theta^i, \theta') = \min(1, p_\theta)$, where

$$p_\theta = \frac{\pi(\theta') f(\mathbf{y}|\theta')}{\pi(\theta^i) f(\mathbf{y}|\theta^i)} \frac{q(\theta^i|\theta')}{q(\theta'|\theta^i)}. \tag{4.23}$$

If the move is accepted, θ^{i+1} is set equal to θ'; else θ^{i+1} is set equal to θ^i.

Beginning with $i = 0$, and an initial arbitrarily chosen value for θ, denoted θ^0, a sequence of θ's is generated for $i = 1, \ldots, B + N$. The first B iterations are the so-called burn-in period which is the period prior to convergence of the sample to the desired limiting distribution, and the next N iterations are the inference sample.

In the more general case, where θ is a vector, the above Metropolis–Hastings algorithm generalises immediately, where the proposal density q is a multivariate pdf. An alternative approach (and probably the most common approach in general) is the single-update Metropolis–Hastings algorithm. In this algorithm, each iteration of the Metropolis–Hastings step involves cycling through each parameter in turn and proposing a new candidate value for the given parameter being updated. This value is accepted with the probability given above (though algebraic simplifications can often be made). A single iteration is completed after each parameter has been updated. An example of the single-update algorithm is described mathematically in Sect. 4.5.1.2.

4.5.1.1 Implementation Issues

There are several decisions that must be made when carrying out MCMC simulations using the Metropolis–Hastings algorithm. Here we just list some of the issues; Gilks et al. (1996) and Brooks et al. (2011) are excellent guides to practical implementation.

1. *Burn-in length, B*. There are various ways to identify a suitable value for B. Usually, time series plots of the iterated values, also known as trace plots, are visually examined to see whether or not the simulated values are changing relatively rapidly as i increases. This is visual evidence for what is termed "good mixing", i.e. relatively rapid sampling of the posterior distribution with little "sticking" at particular values. For example, the simulated values of θ may be generally increasing from 0.1 to 0.4 for $i = 1, \ldots, 100$, and then for $i > 100$, the values are on average 0.4 but vary from 0.2 to 0.6 with no recognisable pattern. In that case, B might be set at 100.

   A less subjective approach is to start the simulation several times, i.e. to run multiple chains, each time using a different initial value, and then to examine the point at which the chains overlap consistently. A formal measure of convergence is the Brooks-Gelman-Rubin (B-G-R) statistic (Brooks and Gelman 1998), which is essentially an F-ratio statistic that compares the variation between chains with variation within chains. Values of the B-G-R statistic near 1.0 are considered indicative of convergence.

2. *Chain length, N*. The sample size can be chosen pragmatically on the basis of the Monte Carlo error of statistics of the posterior sampling distribution. For example, multiple chains are run and the 0.05 percentiles for θ are compared between chains, and if the between-chain variation, i.e. Monte Carlo variation, is considered low enough, then N is sufficient.
3. *Choice of proposal distribution, $q(\theta'|\theta_i)$*. The proposal distribution affects the mixing of the chain and time needed until convergence. There are various somewhat mechanical choices for proposals. In some cases, the full conditional distributions, e.g. $\pi(\theta_1|\mathbf{y}, \theta_2, \ldots, \theta_q)$, are known and these conditionals then serve as proposals. In this case the acceptance probabilities of Equation (4.23) are 100 %, i.e. every candidate value is kept, and the Metropolis–Hastings sampler becomes what is known as the Gibbs sampler. For other alternatives including random walk proposals and independence proposals, see Gilks et al. (1996).
4. *Blocking*. With multiple parameters, e.g. $\theta = (\beta_0, \beta_1, \sigma^2)'$ as in a simple linear regression, we can choose to implement a single-update Metropolis–Hastings algorithm or simultaneously generate candidate values for two or more parameters at one time. Simultaneously updating two or more parameters is known as blocking and is most commonly used when the parameters within a block are highly correlated as it can improve mixing (allowing a smaller value of N and reducing the number of iterations B until convergence).

4.5.1.2 MCMC for SSMs

Here we describe one way that the Metropolis–Hastings algorithm can be used to generate a sample from the joint posterior distribution of states and parameters, $\pi(\mathbf{n}_{0:T}, \boldsymbol{\theta}, \boldsymbol{\psi}|\mathbf{y}_{1:T})$. For an early application of MCMC methods, in particular the Gibbs Sampler, for a non-normal, nonlinear SSM see Carlin et al. (1992).

To begin, prior distributions $\pi(\boldsymbol{\theta}, \boldsymbol{\psi})$ and $g_0(\mathbf{n}_0|\boldsymbol{\theta})$ for the parameters and the initial state vector are specified. Here we assume that a full iteration of the sampler produces a complete vector of all the state values and parameters; i.e. at the end of iteration i, all the states and parameters are "updated". Furthermore, at each iteration assume that the chain first produces individual components of the parameter vector one at a time, then the state vectors, $\mathbf{n}_0$, $\mathbf{n}_1$, ..., $\mathbf{n}_T$ are generated in sequence, and assume that all components of the state vector at time t are produced simultaneously.

For convenience, we denote the combined vector of state and observation parameters by $\boldsymbol{\eta} = \{\boldsymbol{\theta}, \boldsymbol{\psi}\}$ with elements η_r, $r = 1, \ldots, R$. At iteration i, we cycle through each element of $\boldsymbol{\eta}$ in turn, propose a candidate value for the given parameter, and accept/reject this candidate value with the given specified acceptance probability. Mathematically, suppose that we propose to update parameter η_r, such that (for $r > 1$) we have already updated parameters $\eta_1, \ldots, \eta_{r-1}$. We denote the set of current parameters by $\boldsymbol{\eta}^i_{[r]} = \{\eta^{i+1}_1, \ldots, \eta^{i+1}_{r-1}, \eta^i_r, \eta^i_{r+1}, \ldots, \eta^i_R\}$ and current state vectors, $\mathbf{n}^i_{0:T}$. Simulate a *candidate* value, denoted η'_r from the *proposal* distribution with pdf $q(\eta'_r|\eta^i_r)$, and set $\boldsymbol{\eta}'_{[r]} = \{\eta^{i+1}_1, \ldots, \eta^{i+1}_{r-1}, \eta'_r, \eta^i_{r+1}, \ldots, \eta^i_R\}$. The candidate value, η'_r is accepted with probability $\alpha_{\eta_r} = \min\left(1, p_{\eta_r}\right)$, where

$$p_{\eta_r} = \frac{\pi(\boldsymbol{\eta}'_{[r]}) g_0(\mathbf{n}^i_0|\boldsymbol{\eta}'_{[r]}) \prod_{t=1}^T f_t(\mathbf{y}_t|\mathbf{n}^i_t, \boldsymbol{\eta}'_{[r]}) g_t(\mathbf{n}^i_t|\mathbf{n}^i_{t-1}, \boldsymbol{\eta}'_{[r]})}{\pi(\boldsymbol{\eta}^i_{[r]}) g_0(\mathbf{n}^i_0|\boldsymbol{\eta}^i_{[r]}) \prod_{t=1}^T f_t(\mathbf{y}_t|\mathbf{n}^i_t, \boldsymbol{\eta}^i_{[r]}) g_t(\mathbf{n}^i_t|\mathbf{n}^i_{t-1}, \boldsymbol{\eta}^i_{[r]})} \frac{q(\eta^i_r|\boldsymbol{\eta}'_{[r]})}{q(\eta'_{[r]}|\eta^i_{[r]})}.$$

If the move is accepted, set $\eta^{i+1}_r = \eta'_r$; otherwise, set $\eta^{i+1}_r = \eta^i_r$. Following completion of each parameter update, we denote the set of updated parameters by $\boldsymbol{\eta}^{i+1}$.

Similarly for the (block) updating of the states, we cycle through each state vector in turn ($\mathbf{n}_t$ for $t = 0, \ldots, T$), simulate a candidate state vector, and accept or reject the proposed values. Consider time $t = 1, \ldots, T-1$, with current set of all state vectors, $\mathbf{n}^{i+1}_0, \ldots, \mathbf{n}^{i+1}_{t-1}, \mathbf{n}^i_t, \mathbf{n}^i_{t+1}, \ldots, \mathbf{n}^i_T$. Propose a candidate state vector for time t, denoted $\mathbf{n}'_t$, from some proposal density $q_{\mathbf{n}_t}(\mathbf{n}'_t|\mathbf{n}^i_t)$. The proposed state vector is accepted with probability $\alpha_{\mathbf{n}_t}(\mathbf{n}^i_t, \mathbf{n}'_t) = \min(1, p_{\mathbf{n}_t})$, where

$$p_{\mathbf{n}_t} = \frac{f_t(\mathbf{y}_t|\mathbf{n}'_t, \boldsymbol{\theta}^{i+1}) g_t(\mathbf{n}'_t|\mathbf{n}^{i+1}_{t-1}, \boldsymbol{\psi}^{i+1}) g_{t+1}(\mathbf{n}^i_{t+1}|\mathbf{n}'_t, \boldsymbol{\psi}^{i+1})}{f_t(\mathbf{y}_t|\mathbf{n}^i_t, \boldsymbol{\theta}^{i+1}) g_t(\mathbf{n}^i_t|\mathbf{n}^{i+1}_{t-1}, \boldsymbol{\psi}^{i+1}) g_{t+1}(\mathbf{n}^i_{t+1}|\mathbf{n}^i_t, \boldsymbol{\psi}^{i+1})} \frac{q_{\mathbf{n}_t}(\mathbf{n}^i_t|\mathbf{n}'_t)}{q_{\mathbf{n}_t}(\mathbf{n}'_t|\mathbf{n}^i_t)}$$

(explicitly using the terms $\boldsymbol{\theta}$ and $\boldsymbol{\psi}$ instead of the $\boldsymbol{\eta}$ terminology above). If the move is accepted, set $\mathbf{n}^{i+1}_t = \mathbf{n}'_t$; else set $\mathbf{n}^{i+1}_t = \mathbf{n}^i_t$. We note that for the cases $t = 0$ and $t = T$, the acceptance probability simplifies further. For $t = 0$, the term $g_t(\mathbf{n}_t|\mathbf{n}_{t-1}, \boldsymbol{\psi}) = g_0(\mathbf{n}_0|\boldsymbol{\psi})$; for $t = T$, the term $g_{t+1}(\mathbf{n}_{t+1}|\mathbf{n}_t, \boldsymbol{\psi}) \equiv 1$.

In practice, efficient implementation of an MCMC sampler for a SSM can be very difficult (Newman et al. 2009; Fearnhead 2011; King 2011). For example, correlation between states and parameters can be very high which means that very long simulations of the Markov chain may be necessary before the chain converges to the posterior distribution.

The resulting posterior sample of states is a sample from the *smoothed* distribution, i.e. all observations, $\mathbf{y}_{1:T}$, are used to make inferences about states at all time periods. This is in contrast to some inference procedures for states which produce samples from the *filtered* distribution, i.e. only observations up to and including time t, $\mathbf{y}_{1:t}$, are used to make inferences about $\mathbf{n}_t$.

4.5.2 BUGS Project Software

The BUGS (Bayesian inference Using Gibbs Sampling) project (www.mrc-bsu.cam.ac.uk/bugs/) has led to the freely available software WinBUGS and OpenBUGS for model fitting using MCMC (Lunn et al. 2000, 2013). WinBUGS is a stable version of BUGS (the final version of this software is 1.4.3) while OpenBUGS is an open-source version "on which all future development work will be focused". As suggested by the name, WinBUGS is restricted to Windows operating systems (or Windows emulators), while OpenBUGS is available for Windows and Linux (using a command line interface). These packages have a graphical user interface (GUI) but can also be run from within `R` using the associated packages `R2WinBUGS` (for interacting with `WinBUGS`) and `R2OpenBUGS` or `BRugs` (for interacting with `OpenBUGS`). The BUGS language is also used in the software `JAGS` (Plummer 2003), freely available from http://mcmc-jags.sourceforge.net with versions for Windows, Linux and Mac OS X operating systems. JAGS has a command line based interface, but with an associated `R` package `rjags` for calling the program from `R`. For in-depth discussion of the BUGS language, numerous examples and further description of the different packages, including differences between them, see Lunn et al. (2013).

We will demonstrate the use of `WinBUGS` to fit a SSM but before doing so, we give a simpler example of Bayesian inference for a linear regression.

4.5.2.1 Set-Up

There are three sets of program code for most `WinBUGS` programs: (i) the model statement, (ii) the data statement, and (iii) the initial values.

The model statement consists of pieces of code that define the priors and the likelihood. For example, for a simple linear regression, where $y \sim \text{normal}(\beta_0 + \beta_1 x, \sigma^2)$:

```
 #(i) Model Statement
model {
 #Prior distribution for parameters
 beta1 ~ dnorm(0,0.01)
 beta0 ~ dnorm(0,0.01)
 sigma ~ dunif(0.01,10)
 tau   <- 1/(sigma*sigma)  #the precision

 #Likelihood defined
 for(i in 1:n) {
   mu[i] <- beta0+beta1*x[i]
   y[i] ~ dnorm(mu[i],tau)
 }
}
```

Note that the priors for the coefficients, β_0 and β_1, are normal($\mu = 0, \sigma^2 = 1/0.01 = 100$), as the `WinBUGS` syntax for normal random variables specifies the precision, or inverse of the variance. The symbol # denotes a comment in `WinBUGS`.

The data statement specifies the values corresponding to the observations, relevant covariates, and sample size that have been used in the model statement. Again referring to the simple linear regression example:

```
#(ii) Data Statement

list(n=15,
      y=c( 29.4, 33.8, 24.8, 26.1, 31.6, 29.4, 14.6,
        12.6, 29.3, 27.5,28.9, 28.3 ,24.7, 28.5,
        14.9),
      x=c(8, 10,  7,  7,  9, 9,  3,  3,  8,  7,  8,
        8,  7,  8,  3))
```

The initial values are the $\boldsymbol{\theta}^0$s used to begin the chain. e.g.

```
#(iii) Initial Values Statement

list(beta0=0.5, beta1=1.0, sigma=3.0)
```

Note that, in general, it is possible to generate all (or some of) the initial values in `WinBUGS`. In this case the starting values are generated from the corresponding prior distribution specified for the parameters. However, not all prior distributions can be simulated from, in which case an error message is displayed stating that an initial value cannot be generated for the given distribution. To run the MCMC simulations, initial starting values do need to be specified for (at least) these parameters. For example, `WinBUGS` cannot simulate from a $\Gamma(0.001, 0.001)$ prior.

For SSMs, initial values need to be specified (or generated) for all the parameters and unknown states. However, care typically needs to be taken in specifying the initial values for the unknown states so that impossible state transitions (given the initial parameter values) are not specified. In practice, the observation vector $\mathbf{y}_{1:T}$ is often used for the observed parts of the state vector (although this can still lead to impossible state transitions). More generally, given the specified model parameters and initial value for $\mathbf{n}_0$, the remaining initial state vectors $\mathbf{n}_{1:T}$ can be obtained by simulating forward from the given model (see Sect. 4.5.3 for an example).

4.5.2.2 Program Execution

The sequence of steps to run `WinBUGS` and produce a sample is:

1. *Check* model statement for syntax errors.
2. *Load data.*
3. *Compile model* to check for compilation errors.
4. *Set initial values* for $\boldsymbol{\theta}^0$.
5. *Specify model outputs*, i.e. the parameters for which MCMC output is to be shown.
6. *Specify number of iterations.*

More than one chain can be generated, which then allows the B-G-R statistic to be calculated. Note that it is up to the user to specify the burn-in, B, to be used in any posterior summary estimates.

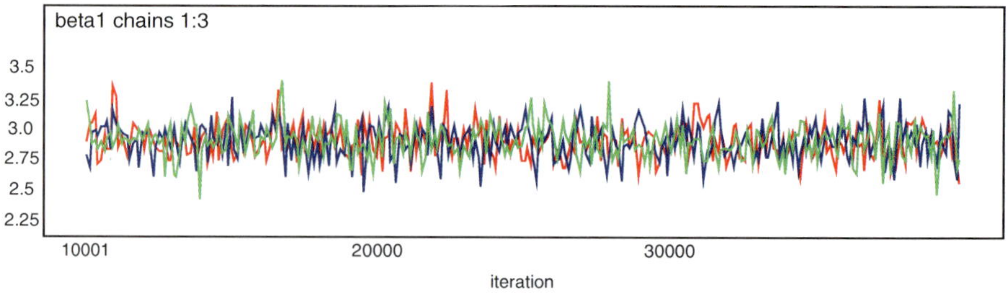

node	mean	sd	MC error	2.5%	median	97.5%
beta0	5.61	1.043	0.03399	3.50	5.56	7.738
Beta1	2.86	0.142	0.00442	2.58	2.86	3.142
Sigma	1.14	0.252	0.00856	0.78	1.10	1.778

Fig. 4.1 `WinBUGS` output from fitting a simple linear regression. The trace plot shown is for the slope coefficient β_1, based on three chains with differing starting values, $B = 10{,}000$ and $N = 30{,}000$

4.5.2.3 Program Output

The output from `WinBUGS` includes trace plots of the simulated parameter values, the B-G-R statistic if multiple chains are run, and summary statistics from the posterior distribution (including the mean, standard deviation, and various quantiles). Recall that the burn-in, B, should be initially determined (for example using the B-G-R statistic) and then discarded before calculating these summary statistics. Example output from fitting the simple linear regression model is shown in Fig. 4.1. Three chains were run with different starting values, each with a burn-in of 10,000 iterations followed by 30,000 iterations for inference. The posterior mean values for β_0, β_1 and σ shown in the figure are quite similar to least squares estimates of 5.69, 2.85 and 1.04. The MC error, Monte Carlo error, is computational error in the calculation of the mean value and can be made arbitrarily small by increasing the number of simulations.

4.5.3 Fitting the Simplified Coho Salmon SSM with `WinBUGS`

For the univariate coho salmon SSM defined previously (Sect. 3.2.1), a Ricker population dynamics model with Poisson variation and bias-corrected lognormal observations, was fitted using `WinBUGS`. Twenty years of observations were used (the first 20 shown in Fig. 3.2; see the `R` code on the book website (Sect. 1.2) for Chap. 3). These had a coefficient of variation of 30 % (`CV.obs`). The model definition code, available on the book website, estimates three parameters, α, β and n_0, assigned uniform prior distributions, while the observation CV was treated as a known constant.

Fig. 4.2 Fitted states, true states and estimates for univariate coho salmon SSM with Ricker population dynamics (Poisson variation) and lognormal observations

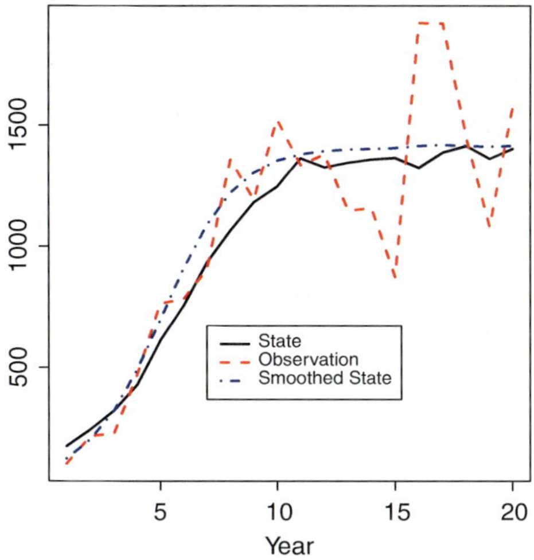

Table 4.2 Posterior means for parameters of state process of the univariate coho salmon model based on MCMC sampling procedure

Parameters	True	SSM	Non-SSM
α	1.50	1.73	1.91
β	3.00e-4	3.90e-4	4.40e-4
n_0	135	76.4	75.1

Means based on the correctly specified SSM are in the column SSM while means based on fitting a non-SSM Ricker model are in the column non-SSM

Rather than inputting the model statement, data and initial values directly within `WinBUGS`, the values were passed from `R` using the function `bugs` from the `R` package `R2WinBUGS`. The relevant `R` code is also available on the book website. This is particularly convenient for generating multiple sets of initial values $\boldsymbol{\theta}_0$ (α, β and $n_{0:20}$) for running multiple MCMC chains. A case-specific function `init.value.generator` was written in this case to provide initial values for the entire state variable sequence.

The posterior means of the state variables are plotted along with the true state values and observations in Fig. 4.2. The posterior means for the parameters α, β and n_0 are shown along with the true values in Table 4.2. For comparison, a non-SSM model was fitted, where the estimated abundances were modelled according to the Ricker model with lognormal errors, i.e.

$$\hat{n}_t|\hat{n}_{t-1} \sim \text{lognormal}\left(\log\left(\alpha\hat{n}_{t-1}\exp(-\beta\hat{n}_{t-1})\right) - \sigma^2/2, \sigma^2\right).$$

Both models overestimated the productivity parameter α and the density dependence parameter β, but the SSM estimates were closer to the true values.

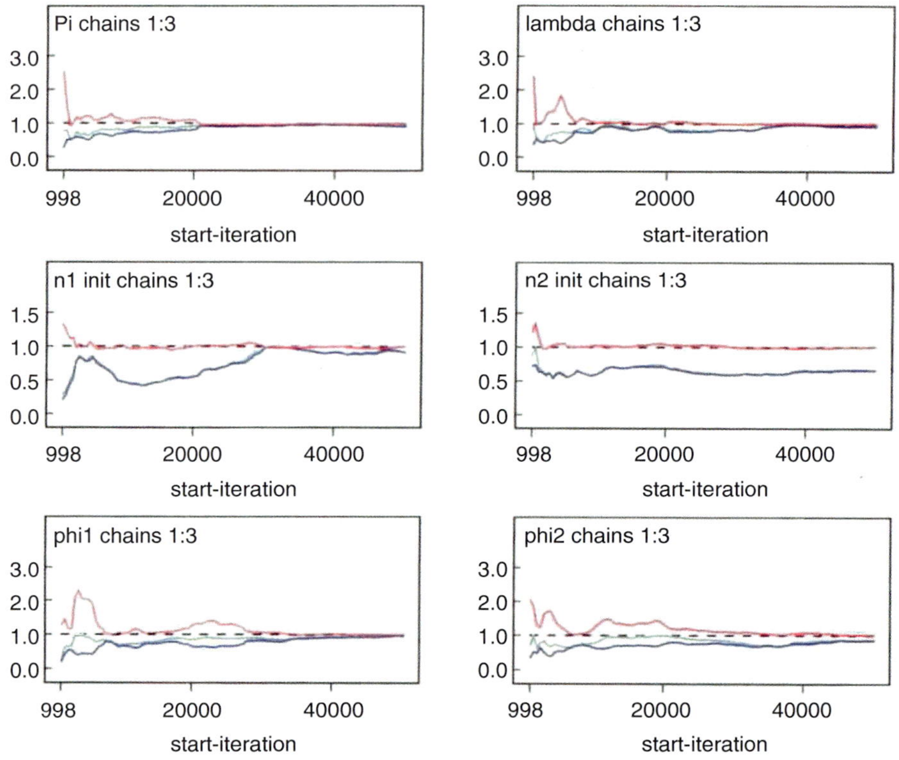

Fig. 4.3 Plot of B-G-R statistic versus iteration based on 3 independent MCMC chains for the parameters π, ρ, $n_{1,0}$, $n_{2,0}$, ϕ_1 and ϕ_2. The *green line* is the width of the central 80 % interval for all three runs pooled and based on bins of length 50, while the *blue line* gives the average width of the 80 % intervals within all three runs individually, and the *red line* is the ratio R of pooled to averaged within credible interval widths. Ideally R converges to 1 and the pooled and within lines converge

4.5.4 *Fitting the BRS SSM with* `WinBUGS`

For a somewhat more complex SSM, the BRS model (Sect. 3.2.2) was fitted using `WinBUGS`. The `WinBUGS` code is available on the book website (Sect. 1.2) as is the generating `R` code. As mentioned previously, with a sequence of three sub-processes, the pdf for $\mathbf{n}_t|\mathbf{n}_{t-1}$ is intractable. The `WinBUGS` code deals with this by explicitly defining the intermediate or "latent" states, the u's, and their corresponding pdf's. Three different chains were run; each had a different set of initial values for the parameters as well as for the latent states, all of which were generated within `R`.

The B-G-R statistic (Fig. 4.3) indicated that the chain was not converging for the initial population size $n_{1,0}$, which is possibly indicative of non-identifiability or weak identifiability problems. However, the posterior mean (smoothed) values for the states matched well with the true states (Fig. 4.4).

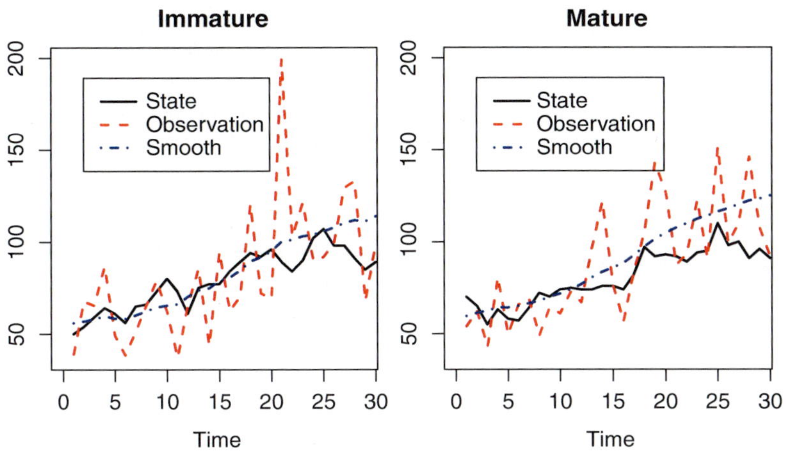

Fig. 4.4 Fitted states, true states and estimates for BRS SSM with lognormal observations

Table 4.3 Parameters and posterior means for the BRS model based on 30 years of simulated data

	True	Posterior distribution		
Parameter	value	2.5 %	Mean	97.5 %
ϕ_1	0.60	0.13	0.50	0.92
ϕ_2	0.70	0.37	0.76	0.94
ρ	0.70	0.37	0.72	0.93
$n_{1,0}$	50	13	89	318
$n_{2,0}$	70	13	54	122

The true parameter values along with posterior distribution summaries, based on the combined output of three MCMC chains of length 50,000 following a burn-in of 50,000, are shown in Table 4.3.

4.5.5 Sequential Importance Sampling

We discuss sequential importance sampling (SIS) in the context of Bayesian inference for a SSM. Once again the objective is to produce a sample from the posterior distribution for the parameters and states. The collection of papers edited by Doucet et al. (2001) is a good reference for details of sequential importance sampling in its various forms.

4.5.5.1 Importance Sampling

Before discussing SIS, we describe "ordinary" importance sampling. Suppose we want a sample of values of a random variable X from a distribution with pdf $f(x)$.

Further suppose that generating samples directly from $f(x)$ is difficult, but given a particular value x, $f(x)$ can be evaluated. Let $g(x)$ be the pdf for a random variable with support including the support of $f(x)$, i.e. if $f(x) > 0$, $g(x) > 0$. This simply means that if $f(x)$ can produce values between 0 and 20, say, then $g(x)$ needs to be able to produce values between 0 and 20, too. The distribution of interest, $f(x)$, is referred to as the target distribution and the generating distribution, $g(x)$, is the trial distribution (Liu 2001).

The following importance sampling algorithm yields an independent sample from $f(x)$:

1. Generate a sample of N values from $g(x)$, $x_1^*, \ldots, x_N^*$.
2. For each generated value, calculate the ratio, $w'(x_i^*)=f(x_i^*)/g(x_i^*)$.
3. Scale the ratios, $w'(x_i^*)$, so that they sum to 1:

$$w(x_i^*) = \frac{w'(x_i^*)}{\sum_{i=1}^{N} w'(x_i^*)}.$$

4. Resample the x^* by sampling values x_i^* with replacement and with probability $w(x_i^*)$.

We note that, with this particular implementation of an importance algorithm, $f(x)$ need only be evaluated up to an unknown constant of proportionality. The rescaling has the effect of cancelling out the unknown proportionality constant. Thus, as with MCMC, this is particularly relevant for Bayesian inference in that the normalizing constant (the denominator in the posterior pdf, Eq. (4.5)) need not be calculated.

4.5.5.2 Sequential Importance Sampling for SSMs

To use ordinary importance sampling to generate the posterior sample for the states and parameters of an SSM, we need to specify a relatively high-dimensional pdf $g(\mathbf{n}_0, \ldots, \mathbf{n}_T, \boldsymbol{\theta})$. This has sometimes been done using distributions such as a multivariate t-distribution (e.g. Cunningham 2002). An alternative, "divide and conquer" approach is sequential importance sampling (Liu and Chen 1998), which is, as the name suggests, a sequential implementation of importance sampling. Doucet et al. (2001) give a nice description of the method and our explanation below largely follows theirs.

We will just explain how SIS proceeds for an SSM where the parameters and initial state $\mathbf{n}_0$ are known. Dual inference, about the unknown parameters as well as the states, can be done using a method developed by Liu and West (2001).

At time $t = 1$, generate $\mathbf{n}_1^*$ from a trial distribution pdf $h_1(\mathbf{n}_1)$ where h_1 can depend upon $\mathbf{n}_0$ and $\boldsymbol{\theta}$. Evaluate the following weight:

$$w_1 = \frac{g_1(\mathbf{n}_1^*|\mathbf{n}_0, \boldsymbol{\theta}) f_1(\mathbf{y}_1|\mathbf{n}_1^*, \boldsymbol{\psi})}{h_1(\mathbf{n}_1^*)}.$$

Next, generate $\mathbf{n}_2^*$ from a pdf $h_2(\mathbf{n}_2)$ where h_2 can depend upon parameters, previous states, and previous and current observations. Evaluate a new weight

$$w_2 = w_1 \times \frac{g_2(\mathbf{n}_2^*|\mathbf{n}_1^*, \boldsymbol{\theta}) f_2(\mathbf{y}_2|\mathbf{n}_2^*, \boldsymbol{\psi})}{h_2(\mathbf{n}_2^*)}.$$

Proceed in a similar manner for the remaining time periods. After the last time period,

$$\begin{aligned} w_T &= \prod_{t=1}^{T-1} w_t \times \frac{g_T(\mathbf{n}_T^*|\mathbf{n}_{T-1}^*, \boldsymbol{\theta}) f_T(\mathbf{y}_T|\mathbf{n}_T^*, \boldsymbol{\psi})}{h_T(\mathbf{n}_T^*)} \\ &= \prod_{t=1}^{T} \frac{g_t(\mathbf{n}_t^*|\mathbf{n}_{t-1}^*, \boldsymbol{\theta}) f_t(\mathbf{y}_t|\mathbf{n}_t^*, \boldsymbol{\psi})}{h_t(\mathbf{n}_t^*)}. \end{aligned}$$

Note that a special case of a trial distribution is where h_t is g_t, the state process pdf. In this case, the weights are proportional to the observation process pdf, $f_t(\mathbf{y}_t|\mathbf{n}_t^*)$.

The weights are subsequently scaled to cancel out the proportionality constant $1/f(\mathbf{y}_1, \ldots, \mathbf{y}_T)$, and the generated sequences can be re-sampled according to these scaled weights to yield a sample from the posterior distribution. An alternative to accumulating the weights in this manner is to re-sample at various points in time, with probabilities proportional to the current weights, according to some re-sampling schedule. One extreme schedule is to re-sample every time period and eliminate the need to store weights. Such re-sampling schedules can affect the degree of Monte Carlo variation in the posterior samples.

In the case of unknown parameters and initial state $\mathbf{n}_0$, we generate a sample of parameters and initial states from a sampler with joint pdf $h_{\boldsymbol{\theta},\boldsymbol{\psi}} \times h_0$, say. The weight at time T is then

$$w_T = \frac{\pi(\boldsymbol{\theta}, \boldsymbol{\psi}) g_0(\mathbf{n}_0|\boldsymbol{\theta})}{h_{\boldsymbol{\theta},\boldsymbol{\psi}}(\boldsymbol{\theta}, \boldsymbol{\psi}) h_0(\mathbf{n}_0|\boldsymbol{\theta})} \prod_{t=1}^{T} \frac{g_t(\mathbf{n}_t^*|\mathbf{n}_{t-1}^*, \boldsymbol{\theta}) f_t(\mathbf{y}_t|\mathbf{n}_t^*, \boldsymbol{\psi})}{h_t(\mathbf{n}_t^*)}.$$

As for MCMC, efficient implementation of SIS is not necessarily simple. One of the problems is that the distribution of weights can be quite asymmetric with relatively few sample vectors having most of the weight, a problem known as particle depletion. This can be problematic even with very large initial sample size. There are numerous techniques aimed at reducing particle depletion, such as the auxiliary particle filter (Pitt and Shephard 1999; Liu and West 2001), residual resampling (Liu and Chen 1998), and partial rejection control (Liu 2001).

The algorithm of Liu and West (2001) adds the step of kernel smoothing the selected parameter values, after an importance sampling step, to lessen the degree of particle depletion. `R` code implementing this algorithm for the univariate coho salmon SSM is available on the book website (Sect. 1.2). The program was run

Table 4.4 Posterior means for parameters of state process of the univariate coho salmon model based on both MCMC and SIS sampling procedures

Parameters	True	MCMC	SIS
α	1.50	1.73	1.64
β	3.00e-4	3.90e-4	3.54e-4
n_0	135	76.4	104.6

with 200,000 particles, where the initial parameter values were drawn from the same prior distributions used in the MCMC implementation. The degree of kernel smoothing was set at 5 %, i.e. 95 % of the weighting in the smooth came from the original, unsmoothed value. Particle depletion was 99.6 %; i.e. of the 200,000 original particles, only 850 of the original simulated values (ignoring effects of kernel smoothing on parameters which perturbs *all* parameters) were left by the 20th year of simulations. In this particular instance, the SIS posterior means for α, β and n_0 (Table 4.4) were closer to the true values than the MCMC values, but this should not be overstated due to Monte Carlo variation and perhaps due to the bias introduced by kernel smoothing (Newman et al. 2009). The prior and posterior densities for the parameters are shown in Fig. 4.5.

4.6 Selecting a Fitting Procedure

We have discussed several approaches for fitting state-space models, distinguishing classical and Bayesian frameworks as well as analytical and simulation-based solutions. How does one determine which fitting procedure to use? No single procedure is consistently the best choice; e.g. Monte Carlo solutions are not necessarily better than analytic solutions. To make a decision, consideration of the following criteria may be useful.

1. Code development time or ease of implementation.
2. Computational efficiency, namely the computing time to produce parameter and state estimates.
3. Accuracy of estimated parameters and states.
4. Degree of fidelity to reality.
5. Risk associated with decisions made using the SSM.

Newman et al. (2009) compared MCMC and SIS using the first three criteria and concluded that there was no clearcut choice between the two approaches as these criteria are affected by the SSM formulation and the available data. They noted that MCMC and SIS are similar in that one must choose a "sample generating" distribution (a proposal distribution for MCMC and an importance distribution

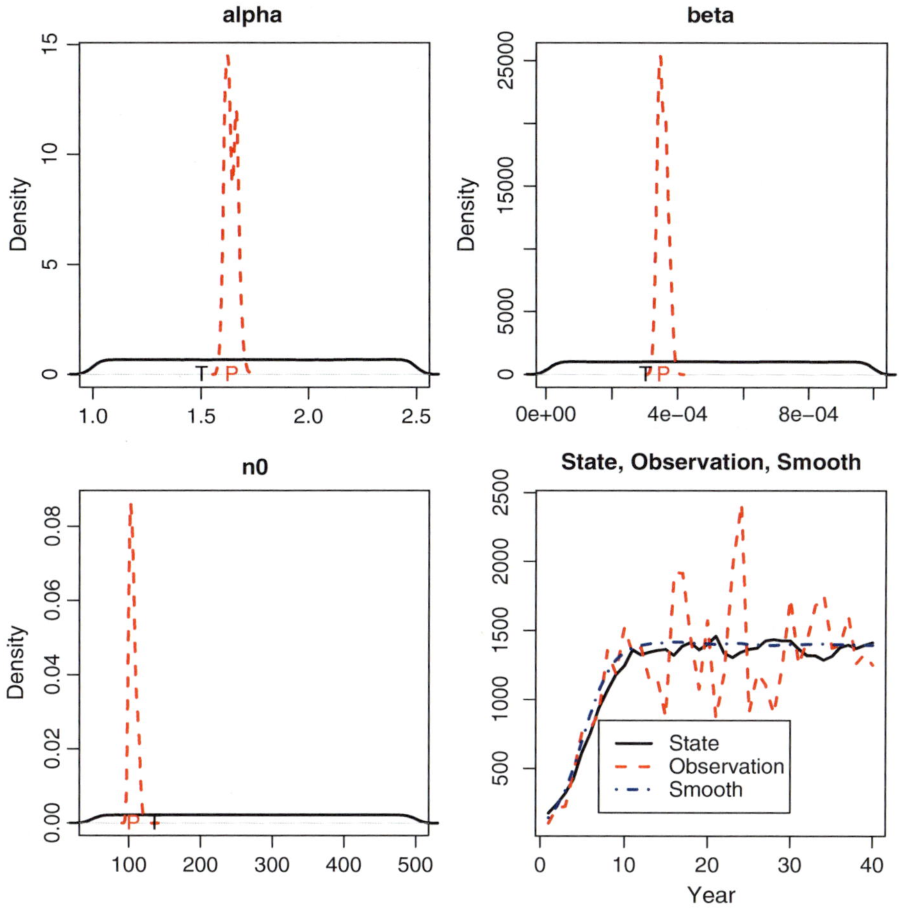

Fig. 4.5 Results from the SIS fit to the univariate coho salmon SSM with Ricker population dynamics (Poisson variation) and lognormal observations. Prior (*solid*) and posterior (*dashed*) for the parameters α, β and n_0 are shown along with marks "T" at the true values (1.50, 3.0e-4, 135) and "P" at the posterior mean values (1.64, 3.5e-4, 105). The *lower right plot* shows the fitted states, true states and estimates

for SIS), and those choices can affect accuracy of estimates as well as ease of implementation.

The fourth criterion, fidelity to reality, refers to how similar the postulated SSM is thought to be to reality. There is an interplay between selecting a model-fitting procedure and model formulation. If time to develop and execute computer code is quite limited and if one is willing to approximate a nonlinear, non-normal SSM, considered relatively realistic, with a normal dynamic linear model, considered less realistic, then maximum likelihood estimation using the Kalman filter to calculate the likelihood may be a good choice.

If there is model uncertainty and the set of models to fit is relatively large, then run execution time can be a deciding criterion and analytic fitting procedures may be preferable.

The fifth criterion, risk associated with use of the SSM, can be related to accuracy and model fidelity. If the consequences of poor decisions resulting from using a less-than-adequate SSM are high, then model fidelity and accuracy of estimates can be extremely important. Devoting considerable financial resources to code development and computational time may be preferable to the potential risk of poor management decisions. If a relatively complex realistic SSM is considered necessary, then Monte Carlo techniques may be the only choice.

The length of the time series, the length of the state vector, and the length of the observation vector, as well as the nature of the relationship between observation vector components and state vector components, can all affect accuracy of inferences about parameters and states. As time series length increases, particle depletion in some sequential Monte Carlo procedures in particular can reduce accuracy.

Issues pertinent to model formulation are discussed further in Sect. 5.1.

Chapter 5
Model Formulation and Evaluation

The previous two chapters have presented the state-space model as a general framework for modelling population dynamics and discussed alternative ways of fitting SSMs to data. In this chapter, we address model formulation and model evaluation.

How does one formulate a model, or models, for an animal population in the first place? Model construction was considered from the perspective of using matrix models as building blocks in Chap. 2. In Sect. 5.1 we take a more general look at the issue. The initial formulation of a model may be one that cannot be practically fit to data, in the sense that model parameters and state vector values cannot be estimated. Even if the model is biologically sensible, the available data might be inadequate, e.g. the state vector consists of age classes $0, 1, \ldots, 9, 10+$, but data are only available for age 0 and ages $1+$. Or there are too many intermediate sub-processes given the temporal resolution of the state vector. In Sect. 5.2, a general approach to the issue of determining model "over-parameterization", which we label parameter redundancy, is described. In Sect. 5.3, we demonstrate how parameter redundancy can be determined in state-space models.

Once one or more models have been formulated and fitted to the data, how do we determine if a model is adequate, or if one model is better than another? We consider several facets of model evaluation including model comparison and selection (Sect. 5.4), model averaging (Sect. 5.5), and individual model diagnostics (Sect. 5.6).

5.1 Model Formulation

Before formulating a model, and attempting to fit it, the purpose of modelling needs to be clearly articulated. Starfield and Bleloch (1991) generally advise that a model should be formulated to answer one or more specific questions, and to make the model no more complex than necessary. Questions thus motivate and guide

K.B. Newman et al., *Modelling Population Dynamics: Model Formulation, Fitting and Assessment using State-Space Methods*, Methods in Statistical Ecology,
DOI 10.1007/978-1-4939-0977-3_5,

the modelling effort, and these questions typically involve estimation of states or parameters. A progression of questions parallels progression in the understanding of the biology of the animal population. The specificity of the questions provides guidance about the required level of detail and complexity of models. To begin, we might simply ask

- For years $t = 1, \ldots, T$, what were the total abundances N_t?

A follow-up question might focus on a partitioning or categorization of abundances:

- For each year t, what were the abundances for age class a and sex s in location r, $n_{a,s,r,t}$?

At a slightly deeper level, we might want to know the vital rates that led to these abundances:

- For each year, what were the survival, birth, immigration and emigration rates?

Digging deeper, we might ask what factors led to, or influenced, the particular values of the vital rates:

- Did environmental covariates or resource management actions, quantified by variates $x_{j,t}$, affect any or all of the survival, birth, immigration and emigration rates for years $1, 2, \ldots, T$? For example, in the following model for survival,

$$\phi_{j,t} = \frac{1}{1 + \exp\left(\beta_0 + \beta_1 x_{j,t}\right)},$$

  what is the value of β_1?

Modelling objectives will often also include prediction, and with questions paralleling many of the above; e.g. "What will the population abundances be over the next five years?", or "If this management action is taken, how will birth rates be affected?".

Clarity about modelling objectives assists model formulation. We advocate beginning with the state process model, aiming to align its structure to meet the modelling objectives, deferring consideration of the observation model and the available data. Think about what is wanted first, worry about whether it is actually achievable second. The modelling objectives can provide guidance in formulation of the state process model in the following respects:

1. spatial and temporal scope (e.g. geographic extent and number of years of interest);
2. temporal resolution (e.g. abundances on a monthly basis);
3. spatial resolution (e.g. abundances distinguished by habitat type);
4. biological "resolution" of the population (e.g. categorization by age, sex, maturity);
5. specification of the processes which need explicit inclusion (e.g. for fish, separate the survival by life stage: egg, fry, juveniles, etc.).

The building block matrix model formulations, as discussed in Chap. 2, are a useful initial approximation to formulating the state process model. The temporal, spatial and biological resolution desired or needed will be reflected by the definition of the components of the state vector, e.g.

$$n_t \text{ or } \begin{bmatrix} n_{0,t} \\ n_{1+,t} \end{bmatrix} \text{ or } \begin{bmatrix} n_{0,male,t} \\ n_{0,female,t} \\ n_{1+,male,t} \\ n_{1+,female,t} \end{bmatrix} \text{ or } \begin{bmatrix} n_{0,male,south,t} \\ n_{0,female,south,t} \\ n_{0,male,north,t} \\ n_{0,female,north,t} \\ n_{1+,male,south,t} \\ n_{1+,female,south,t} \\ n_{1+,male,north,t} \\ n_{1+,female,north,t} \end{bmatrix}.$$

Specification of the processes driving the dynamics will be approximated by the projection matrices, with multiple sub-processes translating into multiple matrices. In many cases, the general sub-processes will include survival, birth, growth and movement, where survival sub-processes might be further partitioned by natural and anthropogenic sources of mortality.

Using the matrices only as an approximate model blueprint, more realistic state process models, namely the state pdf's for a state-space model (Chap. 3), are next specified. Probabilistic formulations of the sub-processes represented by different matrices composing the projection matrices are translated into multiple pdf's for each sub-process; e.g. binomial distributions for survival, Poisson distributions for births, multinomial distributions for movement.

The imaginative, and often idealized, state process model formulation is brought back to reality at this point by asking:

> If we had perfect information about every component in the state vectors over T time periods, could the parameters of the state process be estimated?

In the next two sections, we discuss approaches for trying to answer this question, but for now assume that the answer is "No". Given this answer, we might either reformulate (simplify) the state processes, or we might change the resolution of the state vector, e.g. adding state vectors for time periods immediately following a given sub-process might lead to estimability in terms of known state vectors.

Once we have a state process model with state vectors adequate for estimating the unknown process parameters, additional reality is imposed by considering the available data and corresponding observation models, i.e. by formulating observation process pdf's. Often a decision must be made between letting the observations be the data at their most raw level, completely unsummarized, versus using quantities derived from the sample data, e.g. mark-recapture point estimates and associated standard errors. Incorporating raw data into the observation model can be somewhat technically difficult or cumbersome. For example occupancy data, mark-recapture data or distance sampling data are often analysed with sophisticated specialized software, rather than simply using summarized estimates. Recent work by Knape et al. (2013) shows that in some cases inferences for SSMs using the raw data and the summarized data are quite similar. Whether working with raw or summarized data, however, a key question is "given the available data (and observation pdf's), can the state process model parameters and states be estimated?".

Estimability problems will often become obvious during model fitting. Using frequentist methods to fit a SSM, estimability problems may be detected when the optimization program fails to converge. For Bayesian inference, such problems might be detected because posterior distributions are essentially the same as prior distributions, or because of strong dependencies in the posterior distribution, e.g. the joint posterior distribution for two parameters is tightly concentrated along a line or curve. Other differences between frequentist and Bayesian approaches with regard to estimability are noted in Sect. 5.2, which presents general procedures for determining estimability issues in advance of model fitting.

Once an estimability problem has been detected, there are several alternatives. One is to formulate a new SSM; this could be a shortened state vector, with dimension and components more closely matching that of the observation vector, or it could be a simpler state process, with fewer sub-processes and corresponding parameters. Alternatively, we might keep the model as is, but then "arbitrarily" specify fixed values for some of the unknown parameters. In the case of Bayesian inference, we might specify relatively informative prior distributions for some parameters. If relevant data inform the priors, then the use of such priors will be less questionable than when relevant data are lacking and expert knowledge is used instead to specify the priors. In the longer term, estimation problems due to data deficiencies or inadequacies could lead to new types of data being collected, e.g. recording age or sex of animals during surveys, or it could lead to changes in the frequency of sampling, e.g. increasing data collection from once a year to twice a year, once following an annual birth period and once following an annual migration period.

5.2 Model Formulation: Parameter Redundancy

In this section, we discuss a general approach to the problem of detecting, and possibly rectifying, estimability problems with statistical models. The approach is an area of active research,[1] and the underlying principles are useful for model builders to understand. The perspective taken is that of estimation using frequentist or "classical" methodology, although for a Bayesian approach, problems with the likelihood often translate directly into problems with the posterior distribution, depending upon the nature of the prior distribution.

---

[1]See, for example, http://www.kent.ac.uk/ims/personal/djc24/parameterredundancy.htm.

We begin with a description of the basic ideas of parameter redundancy, starting with historical background. We then explain how computational symbolic algebra methods can be used to determine whether a model is parameter-redundant or not, and if it is, what can be estimated. We introduce the idea of model extensions, in which parameter redundancy results may be extended to cover models that have a fixed structure but are applied to studies of different sizes. An example of such models is the Cormack-Jolly-Seber model (Cormack 1964; Jolly 1965; Seber 1965), where the structure (multinomial models with outcome probabilities being products of time-specific survival and capture probabilities) is the same for T years of capture and recapture, whether $T = 4$ or $T = 7$. We explain how reparameterisation can be used to try to overcome complexity issues. We also consider near-redundant models and connections with Bayesian methods. Then in Sect. 5.3 we show how the approach can be applied for state-space models.

5.2.1 Concepts

A model is parameter-redundant if it is not possible to estimate all of its parameters using classical inference, irrespective of how many data are collected. This may sound peculiar, but a simple illustration is provided in Example 2 below, in which the Cormack-Jolly-Seber model is used to estimate survival from capture-recapture models, in which there is time-dependence of both survival and capture probabilities, as mentioned above. In this example, the last (in time) survival probability and the corresponding last capture probability are confounded; they only appear in the likelihood as a product, and while it is possible to estimate the product (and all the other model parameters) by the method of maximum likelihood, it is not possible to estimate the individual components of the product. Users of the model recognise this deficiency and can live with it. However this example is deceptively simple. Consider a model for recovery data on dead animals. If there is full age-dependence of survival probability and constant probability of recovery and reporting of dead animals (a model called the Seber model, Seber 1971), then once again the model is parameter-redundant. The reason in this case is not self-evident, and indeed it is not possible to estimate any of the original model parameters. Furthermore, intuition is not a reliable guide to parameter redundancy; for instance, if the first-year survival probability in the Seber model is made time-dependent, then the resulting model is no longer parameter-redundant. In this case adding parameters to a redundant model removes the redundancy. Thus we can appreciate that parameter redundancy is all about model structure: how parameters enter into a model, rather than just how many parameters there are.

Adding covariates to a redundant model is another way in which redundancy can be avoided. For example in the Cormack-Jolly-Seber model, one might allow the time-dependent survival probabilities to depend, through logistic regression, on a time-varying covariate; it might describe winter weather, or food availability, for example. The last survival probability then has the logistic regression parameters in common with the other survival probabilities, which potentially allows all parameters in the model to be estimated; see Cole and Morgan (2010b) for more discussion.

Generally, simple models with few parameters are not usually parameter-redundant. Typically, the problems arise when realism and complexity are added. This has been known for many years, in several different areas, such as latent modelling of contingency tables, compartment models, models in econometrics, as well as in models for ecology. A detailed review is provided by Cole et al. (2010), who also describe concepts of parameter identifiability, and how this relates to parameter redundancy. In particular, a parameter-redundant model is non-identifiable and when a model is parameter-redundant, and a likelihood is formed, then the likelihood will not have a unique optimum. For example, Catchpole and Morgan (1991) showed that the Seber model likelihood is maximized along a ridge, and they determined the locus of the ridge in parameter space.

5.2.2 Derivative Matrix and Symbolic Rank

Consider a log-likelihood which is a function of a single parameter, and which is flat for some of the parameter range. Where the log-likelihood is flat, the derivative of the log-likelihood is zero, and the rate of change of the derivative is also zero. This suggests that in general one might detect parameter redundancy by considering the aspects of the Hessian matrix, which is the matrix of second-order derivatives of a log-likelihood surface. Catchpole and Morgan (1997) show that the expected information matrix will be non-singular if and only if a particular derivative matrix **D** has a symbolic rank equal to the number of parameters in the model. This was presented for models that are members of the exponential family, and provides a formal way of testing for parameter redundancy. This result extends the well-known expression of the expected information matrix for the multinomial distribution in terms of a derivative matrix; see Morgan (2009:110). The derivative matrix **D** has rows that correspond to the model parameters, and columns that correspond to the mean values of the observations; the entries of the matrix are the corresponding partial derivatives of means with respect to parameters. The derivative matrix is a function of the model parameters, and so model parameter redundancy can be determined by considering the symbolic rank of the matrix, so that computer packages such as Maple or Mathematica can in theory produce such a symbolic rank. A model is parameter-redundant if its symbolic rank is less than the number of parameters in the model. Two examples are shown below. The first example is a simple closed population, which has a derivative matrix with the same rank as the number of parameters, so that the model is not parameter-redundant. The second example is the time-dependent Cormack-Jolly-Seber model, which is well known to be parameter-redundant, as discussed earlier. Maple code for both these examples is available on the book website (Sect. 1.2).

5.2.2.1 Example 1

A group of n animals is marked and released into a closed population. The population is sampled twice over a relatively short period of time where it is assumed that survival is 100 %. The capture probability at the first time of sampling is p_1 for all animals. Animals not caught the first time are caught the second time with probability p_2. Animals caught the first time, however, experience trap-shyness and they are caught a second time with probability ψp_2, where $0 \leq \psi \leq 1$. Let y_1 be the number of animals caught in the first sampling period but not the second, y_2 the number caught only in the second sampling period, y_3 the number caught in both sampling periods, and $y_4 = n - y_1 - y_2 - y_3$ the number never caught. Then their joint distribution is multinomial:

$$y_1, y_2, y_3, y_4 \sim \text{multinomial}(n, \mathbf{q}),$$

where $\mathbf{q} = \{q_1, q_2, q_3, q_4\} = \{p_1(1-\psi p_2), (1-p_1)p_2, \psi p_1 p_2, 1 - p_1(1-\psi p_2) - (1-p_1)p_2 - \psi p_1 p_2\}$. The expected values of y_1, y_2, y_3 and y_4 form the vector

$$\begin{aligned}\delta &= [nq_1, nq_2, nq_3, nq_4]' \\ &= [np_1(1-\psi p_2), n(1-p_1)p_2, n\psi p_1 p_2, n(1-p_1(1-\psi p_2)-(1-p_1)p_2-\psi p_1 p_2)]',\end{aligned} \tag{5.1}$$

and the parameters form the vector $\boldsymbol{\theta} = [\psi, p_1, p_2]$. The derivative matrix $\mathbf{D}$ is shown below. Columns correspond to the expected values of y_1, y_2, y_3 and y_4; rows correspond to parameters ψ, p_1 and p_2, in that order. Each entry is the partial derivative of the mean with respect to the parameter.

$$\mathbf{D} = \begin{bmatrix} -np_1 p_2 & 0 & np_1 p_2 & 0 \\ n(1-\psi p_2) & -np_2 & n\psi p_2 & n(p_2-1) \\ -np_1\psi & n(1-p_1) & n\psi p_1 & n(p_1-1) \end{bmatrix}. \tag{5.2}$$

The rank of $\mathbf{D}$ is 3. Therefore there is no parameter redundancy and the parameters p_1, p_2 and ψ can be estimated. The maximum likelihood estimates are

$$\hat{p}_1 = \frac{y_1 + y_3}{n} \quad \hat{p}_2 = \frac{y_2}{y_2 + y_4} \quad \hat{\psi} = \frac{y_3(y_2 + y_4)}{y_2(y_1 + y_3)}. \tag{5.3}$$

5.2.2.2 Example 2

Consider the Cormack-Jolly-Seber model used to describe a three-year capture-recapture study. Suppose the data are collated in the form of an m-array with $m_{i,j}$ representing the number of animals released in year i and next captured in year $j+1$ for $1 \leq i \leq 3$ and $i \leq j \leq 3$, and let N_i represent the number of animals released in year i. The corresponding probability that an animal released in year i is first seen in year j is

$$p_{i,j} = \left(\prod_{k=i}^{j} \phi_k\right)\left\{\prod_{k=i+1}^{j} (1-p_k)\right\} p_{j+1} \text{ for } 1 \leq i \leq 3,\ i \leq j \leq 3, \tag{5.4}$$

where $0 \leq \phi_k \leq 1$ is the probability an individual alive at time k survives until year $k+1$, for $k = 1, 2, 3$, and $0 \leq p_k \leq 1$ is the probability of recapture in year k for $k = 2, 3, 4$. Rather than forming a derivative matrix with respect to the mean, Catchpole and Morgan (1997) show that it is sufficient to differentiate the vector consisting of the probabilities $p_{i,j}$,

$$\boldsymbol{\delta} = \begin{bmatrix} \phi_1 p_2 \\ \phi_1\phi_2(1-p_2)p_3 \\ \phi_1\phi_2\phi_3(1-p_2)(1-p_3)p_4 \\ \phi_2 p_3 \\ \phi_2\phi_3(1-p_3)p_4 \\ \phi_3 p_4 \end{bmatrix}, \tag{5.5}$$

with respect to the parameters $\boldsymbol{\theta} = [\phi_1, \phi_2, \phi_3, p_2, p_3, p_4]$. The resulting derivative matrix,

$$\mathbf{D} = \begin{bmatrix} p_2 & \phi_2(1-p_2)p_3 & \phi_2\phi_3(1-p_2)(1-p_3)p_4 & 0 & 0 & 0 \\ 0 & \phi_1(1-p_2)p_3 & \phi_1\phi_3(1-p_2)(1-p_3)p_4 & p_3 & \phi_3(1-p_3)p_4 & 0 \\ 0 & 0 & \phi_1\phi_2(1-p_2)(1-p_3)p_4 & 0 & \phi_2(1-p_3)p_4 & p_4 \\ \phi_1 & -\phi_1\phi_2 p_3 & -\phi_1\phi_2\phi_3(1-p_3)p_4 & 0 & 0 & 0 \\ 0 & \phi_1\phi_2(1-p_2) & -\phi_1\phi_2\phi_3(1-p_2)p_4 & \phi_2 & -\phi_2\phi_3 p_4 & 0 \\ 0 & 0 & \phi_1\phi_2\phi_3(1-p_2)(1-p_3) & 0 & \phi_2\phi_3(1-p_3) & \phi_3 \end{bmatrix}, \tag{5.6}$$

has rank 5. Cole et al. (2010) show that if the derivative matrix is formed from a vector of means then the rank is again 5, and this is also shown in the Maple code for this example. The reason that the ranks are the same is explained in Sect. 5.2.4 below. As this model has six parameters but is of rank of 5, it is parameter-redundant.

The derivative matrix provides more information than whether a model is parameter-redundant or not. In parameter-redundant models, symbolic algebra can also be used to find a set of parameter combinations which can be estimated. In Example 2 there is one non-zero solution to $\boldsymbol{\alpha}'\mathbf{D} = \mathbf{0}$, which is $\boldsymbol{\alpha}' = [0, 0, -\phi_3/p_4, 0, 0, 1]$. The position of the zeros indicates that the parameters ϕ_1, ϕ_2, p_2 and p_3 can still be estimated. In general there will be d non-zero solutions to $\boldsymbol{\alpha}'\mathbf{D} = \mathbf{0}$, where d is the deficiency of the model. The deficiency is calculated as the number of parameters minus the rank of the derivative matrix. If there are zeros in position i of $\boldsymbol{\alpha}$ for all d solutions to $\boldsymbol{\alpha}'\mathbf{D} = \mathbf{0}$, then it is still possible to estimate the ith parameter in $\boldsymbol{\theta}$ even though the model is parameter-redundant. Other combinations of parameters that can be estimated can be found by solving partial differential equations (Catchpole et al. 1998b; Cole et al. 2010). In Example 2, there is one partial differential equation

$$-\frac{\partial f}{\partial \phi_3}\frac{\phi_3}{p_4} + \frac{\partial f}{\partial p_4} = 0, \tag{5.7}$$

the solution to which tells us that the additional parameter that can be estimated is the product, $\phi_3 p_4$. This solution is well-known and obvious, but in less obvious examples Maple code can be used to obtain the estimable parameter combinations.

A model which is not parameter-redundant is said to be full rank, but it may not be full rank for the entire parameter space. This can be investigated through a particular matrix decomposition called the PLUR decomposition or Turing factorization, which can also be produced by Maple; see Cole et al. (2010). It is explained in Cole and Morgan (2010b) and Choquet and Cole (2012) that the PLUR decomposition should be examined when full-rank models contain covariates or non-rational functions of the parameters, to check results and confirm that the model is always full rank. We show in the Maple code for Example 1 that this model is parameter-redundant at boundary values of $p_1 = 0$, $p_1 = 1$ and $p_2 = 0$, and in the unrealistic case of $n = 0$.

5.2.3 General Results

Consider when parameter redundancy is established for a particular size of study, as with a three-year capture-recapture study and a Cormack-Jolly-Seber model. If there were an extra year of recapture, or an extra year of recapture and release, would that change the results? It would be tedious to have to apply the same symbolic algebra calculations each time one has a different length of study for the same structural model. One way to deal with this feature is through the extension theorem of Catchpole and Morgan (1997), which is generalised in Cole et al. (2010). The extension theorem establishes conditions under which parameter redundancy results for a given size of study and for a particular model can be generalized to studies of any size for that type of model, rather like a mathematical proof by induction. In Cole et al. (2010), Example 2 is shown to have rank $2T-3$ for a capture-recapture study of T years with $T-1$ years of capture and $T-1$ subsequent years of recapture. As there are $2T-2$ parameters, this model is always parameter-redundant with deficiency 1 and estimable parameter combinations $\phi_1, \ldots, \phi_{T-2}, p_2, \ldots, p_{T-1}, \phi_{T-1}p_T$. For additional results, see Catchpole and Morgan (2001) and Catchpole et al. (1998a). The papers by Cole et al. (2012) and Hubbard et al. (2014) provide a taxonomy of models for ring-recovery data, capture-recapture data and capture-recapture-recovery data, so that we can check the parameter redundancy status of all the common models, irrespective of study length, without having to do any symbolic algebra. An alternative way to obtain general results is through mathematical theorems for particular models, and examples of these are given by Catchpole et al. (1996), with a focus on age-dependence in models for ring-recovery data, of which the Seber model is just one example.

5.2.4 *Missing Data, Exhaustive Summaries and Reparameterization*

The investigations outlined above are for models, irrespective of the quality of data available. In particular it is assumed that there is no missing information, in that every combination of capture and recapture has occurred. When data points are missing, which is not uncommon in ecological studies, then this has the effect of removing rows from the derivative matrix, as particular means do not enter the likelihood. With fewer rows, the derivative matrix might have smaller rank than previously; although we might know the parameter redundancy status of a model without regard to missing data, any real data set might throw up particular features due to missing data which would mean that we would have to perform the symbolic calculations for that dataset and model combination. An interesting question for any model is how much data can be removed from any study before the parameter redundancy changes; see Cole et al. (2012) for more discussion of this.

A potential problem with using symbolic algebra is that model complexity might mean that computers have insufficient memory to compute symbolic rank. One example of this is described in Cole and Morgan (2010a), for a model for fish survival. They show how appropriate reparameterization simplifies the model, and allows the symbolic algebra to take place. Another example is given in Cole (2012) for multi-state mark-recapture models, in particular those with unobservable states.

The paper by Cole et al. (2010) extends the framework of Catchpole and Morgan (1997), so that more general models than those of the exponential family can be considered. Models are described by exhaustive summaries, which are particular collections of parameters that fully specify models. Several different exhaustive summaries are available for any model. The means for a model form an exhaustive summary and the natural logarithms of the means form an alternative exhaustive summary. The derivative matrices will be different for each different exhaustive summary; however the ranks of each derivative matrix will be the same. In parameter-redundant models, the same solutions will exist to the equation $\boldsymbol{\alpha}^T \mathbf{D} = 0$ and the same set of partial differential equations will result, regardless of the exhaustive summary used.

5.2.4.1 Example 1 Revisited

In Example 1 the multinomial expected values are the exhaustive summary, given by Eq. (5.1). As the last element of $\boldsymbol{\delta}$, δ_4, is a function of δ_1, δ_2 and δ_3 with $\delta_4 = 1 - \delta_1 - \delta_2 - \delta_3$ and because multiplying by the constant n does not change results, a simpler exhaustive summary is

$$\boldsymbol{\delta}_{simp} = [p_1(1 - \psi p_2), (1 - p_1)p_2, p_1 \psi p_2]'. \tag{5.8}$$

The derivative matrix **D** is then

$$\mathbf{D} = \begin{bmatrix} 1 - \psi p_2 & -p_1\psi & -p_1 p_2 \\ -p_2 & 1 - p_1 & 0 \\ \psi p_2 & p_1\psi & p_1 p_2 \end{bmatrix}, \tag{5.9}$$

which also has rank 3.

It can be advantageous using an exhaustive summary which is the natural logarithms of another exhaustive summary if the resulting derivative matrix is simpler. Simpler exhaustive summaries can also be obtained by reparameterizations, and may avoid issues of symbolic algebra packages running out of memory (Cole et al. 2010). Cole (2012) develops a simple exhaustive summary for use with multi-state mark-recapture models and Hubbard et al. (2014) develop a simple exhaustive summary for capture-recapture-recovery models.

5.2.4.2 Example 2 Revisited

For Example 2, rather than use the means, we use the alternative simpler exhaustive summary based on the probabilities given in Eq. (5.5). An even simpler exhaustive summary is given in Cole et al. (2010) as

$$\boldsymbol{\delta} = [\phi_1 p_2, \phi_1(1 - p_2), \phi_2 p_3, \phi_2(1 - p_3), \phi_3 p_4]'. \tag{5.10}$$

The derivative matrix when this exhaustive summary is used is

$$\mathbf{D} = \begin{bmatrix} p_2 & 1 - p_2 & 0 & 0 & 0 \\ 0 & 0 & p_3 & 1 - p_3 & 0 \\ 0 & 0 & 0 & 0 & p_4 \\ \phi_1 & -\phi_1 & 0 & 0 & 0 \\ 0 & 0 & \phi_2 & -\phi_2 & 0 \\ 0 & 0 & 0 & 0 & \phi_3 \end{bmatrix}. \tag{5.11}$$

The rank is again 5 and the solution to $\boldsymbol{\alpha}'\mathbf{D} = 0$ is also $\boldsymbol{\alpha}' = [0, 0, -\phi_3/p_4, 0, 0, 1]$. Therefore the same set of PDEs results to show again that the parameters that can be estimated are $\phi_1, \phi_2, p_2, p_3, \phi_3 p_4$.

5.2.5 *Numerical Methods and Near-Redundancy*

Alternatives to the symbolic approach have been described by Gimenez et al. (2004). We can, for instance, compute an observed information matrix at various points in the parameter space, and each time examine its numerical rank. Various authors have taken this route when straightforward application of symbolic algebra failed due to lack of computer memory; see for example Hunter and Caswell (2009) and Jiang et al. (2007). A hybrid, numerical/symbolic approach is examined by Choquet and Cole (2012). This involves calculation of the derivative matrix symbolically and then evaluating the rank numerically at about five random points in the parameter space. This method is used in the Maple code for Examples 1 and 2 to also obtain ranks of 3 and 5 respectively. Whilst the ranks of the derivative matrix can be obtained in this way as well as by finding solutions to $\boldsymbol{\alpha}^T\mathbf{D} = 0$, it is not possible to find estimable parameter combinations for parameter-redundant models and it is not possible to make use of extension theorems to obtain general results.

Near-redundancy is discussed by Catchpole et al. (2001). It arises when a model is full rank, but as a consequence of particular data, it produces some estimators with low precision, effectively because the model is similar to a sub-model which is parameter-redundant. For example, we might fit a Seber model to ring-recovery data with the addition of time-dependent first-year survival. If the data do not support variation in first-year survival, then the fitted model will be similar to a Seber model, which is parameter-redundant. Near-redundancy might be anticipated from detailed knowledge of the parameter redundancy structure of relevant models, and it can be investigated by the use of numerical methods.

5.2.6 Bayesian Methods

Parameter redundancy is a feature of maximum likelihood. Parameter-redundant models may be fitted using Bayesian inference, with additional information provided by prior distributions. A striking example of this is provided by Brooks et al. (2000) for the Seber model. In this application, where the model is parameter-redundant, marginal parameter precision was high, with narrow credible intervals, even when the prior distributions were flat. This arose due to the orientation of the ridge in the likelihood surface. In general, simply switching methods of inference does not avoid problems with parameter redundancy, as flat likelihood surfaces can result in posterior distributions that cause MCMC samplers to be slow to converge. Uncritical use of Bayesian methods can also result in estimates that are very dependent on prior distributions; see for example Brooks et al. (2000). An analogue of near-redundancy in the Bayesian context arises when there is substantial overlap between a prior distribution and the corresponding posterior. The relevant parameters are said to be weakly identifiable, and illustrations are provided by Gimenez et al. (2009b). One of their examples is the Cormack-Jolly-Seber model.

5.3 Parameter Redundancy in State-Space Models

Before the derivative matrix method can be used to examine parameter redundancy in state-space models, we need a suitable exhaustive summary. Suppose we have a linear state-space model with respective observation and state equations

$$\begin{aligned} \mathbf{y}_t &= \mathbf{O}_t\mathbf{n}_t + \boldsymbol{\epsilon}_t \\ \text{and} \quad \mathbf{n}_t &= \mathbf{L}_t\mathbf{n}_{t-1} + \boldsymbol{\delta}_t. \end{aligned}$$

Cole and McCrea (2012) show that an exhaustive summary is

$$\boldsymbol{\delta} = \begin{bmatrix} \mathrm{E}(\mathbf{y}_1) \\ \mathrm{E}(\mathbf{y}_2) \\ \mathrm{E}(\mathbf{y}_3) \\ \vdots \end{bmatrix} = \begin{bmatrix} \mathbf{O}_1\mathbf{L}_1\mathbf{n}_0 \\ \mathbf{O}_2\mathbf{L}_2\mathbf{L}_1\mathbf{n}_0 \\ \mathbf{O}_3\mathbf{L}_3\mathbf{L}_2\mathbf{L}_1\mathbf{n}_0 \\ \vdots \end{bmatrix}, \tag{5.12}$$

where $\mathbf{n}_0$ are the initial values of the state equations, which can be known constants, parameters to be estimated or any function of the parameters.

5.3.1 Example 3: BRS Matrix Model Example

The BRS matrix model of Eq. (2.11) in Sect. 2.3 has state equation

$$\mathrm{E}(\mathbf{n}_t|\mathbf{n}_{t-1}) = \begin{bmatrix} \mathrm{E}(n_{1,t}) \\ \mathrm{E}(n_{2,t}) \end{bmatrix} = \begin{bmatrix} (1-\pi+\rho\pi)\phi_1 & \rho\phi_2 \\ \pi\phi_1 & \phi_2 \end{bmatrix} \begin{bmatrix} n_{1,t-1} \\ n_{2,t-1} \end{bmatrix}.$$

We assume that both $n_{1,t}$ and $n_{2,t}$ are observed with error so that the observation process is

$$\mathrm{E}(\mathbf{y}_t) = \begin{bmatrix} \mathrm{E}(y_{1,t}) \\ \mathrm{E}(y_{2,t}) \end{bmatrix} = \begin{bmatrix} 1 & 0 \\ 0 & 1 \end{bmatrix} \mathbf{n}_t, \tag{5.13}$$

and also assume the starting values are known constants $n_{0,1}$ and $n_{0,2}$.

The exhaustive summary for this model is

$$\boldsymbol{\delta} = \begin{bmatrix} \begin{bmatrix} \mathrm{E}(y_{1,1}) \\ \mathrm{E}(y_{2,1}) \end{bmatrix} \\ \begin{bmatrix} \mathrm{E}(y_{1,2}) \\ \mathrm{E}(y_{2,2}) \end{bmatrix} \\ \vdots \end{bmatrix} = \begin{bmatrix} \begin{bmatrix} 1 & 0 \\ 0 & 1 \end{bmatrix} \begin{bmatrix} (1-\pi+\rho\pi)\phi_1 & \rho\phi_2 \\ \pi\phi_1 & \phi_2 \end{bmatrix} \begin{bmatrix} n_{0,1} \\ n_{0,2} \end{bmatrix} \\ \begin{bmatrix} 1 & 0 \\ 0 & 1 \end{bmatrix} \begin{bmatrix} (1-\pi+\rho\pi)\phi_1 & \rho\phi_2 \\ \pi\phi_1 & \phi_2 \end{bmatrix}^2 \begin{bmatrix} n_{0,1} \\ n_{0,2} \end{bmatrix} \\ \vdots \end{bmatrix}$$

$$= \begin{bmatrix} (1-\pi+\rho\pi)\phi_1 n_{0,1} + \rho\phi_2 n_{0,2} \\ \pi\phi_1 n_{0,1} + \phi_2 n_{0,2} \\ (1-\pi+\rho\pi)\phi_1\{(1-\pi+\rho\pi)\phi_1 n_{0,1} + \rho\phi_2 n_{0,2}\} + \rho\phi_2(\pi\phi_1 n_{0,1} + \phi_2 n_{0,2}) \\ \pi\phi_1\{(1-\pi+\rho\pi)\phi_1 n_{0,1} + \rho\phi_2 n_{0,2}\} + \phi_2(\pi\phi_1 n_{0,1} + \phi_2 n_{0,2}) \\ \vdots \end{bmatrix}.$$

5.3.2 Using the Exhaustive Summary

There are two things to note about the exhaustive summary in Example 3 above which apply in general to all exhaustive summaries derived using Eq. (5.12). The first is that exhaustive summary terms get increasingly complex for each successive $\mathrm{E}(\mathbf{y}_t)$. The second is that this exhaustive summary is infinite. To use the derivative method in Maple we need a finite exhaustive summary, so that the exhaustive summary is stopped at $\mathrm{E}(\mathbf{y}_\tau)$, for some τ. There are three possible options:

1. If there were T years of data, stop the exhaustive summary at $\mathrm{E}(\mathbf{y}_T)$.
2. If the state-space model is linear with m states and $\mathbf{L}_t = \mathbf{L}$ is not dependent on time, then we only need to consider expansion terms up to $\mathrm{E}(\mathbf{y}_{2m})$.
3. Use the extension theorem of Catchpole and Morgan (1997) and Cole et al. (2010).

Option 1 is only recommended if T is small, because terms get increasingly complex for each successive $\mathrm{E}(\mathbf{y}_t)$. The second option is based on a rule adapted from the Taylor-series method for compartment models (Pohjanpalo 1978; Evans and Chappell 2000) and a proof is given in Cole and McCrea (2012). This rule is demonstrated in Example 3 below. The third option is demonstrated in Example 4 below.

5.3.3 Example 3 Revisited

As there are $m = 2$ states, we can use option 2 and include exhaustive summary terms up to and including $\mathrm{E}(\mathbf{y}_4)$. (Note that we could also use option 3, or option 1 if we had a specific number of years of data.) The exhaustive summary is then

$$\boldsymbol{\delta} = \begin{bmatrix} (1-\pi+\rho\pi)\phi_1 n_{0,1} + \rho\phi_2 n_{0,2} \\ \pi\phi_1 n_{0,1} + \phi_2 n_{0,2} \\ (1-\pi+\rho\pi)\phi_1\{(1-\pi+\rho\pi)\phi_1 n_{0,1} + \rho\phi_2 n_{0,2}\} + \rho\phi_2(\pi\phi_1 n_{0,1} + \phi_2 n_{0,2}) \\ \pi\phi_1\{(1-\pi+\rho\pi)\phi_1 n_{0,1} + \rho\phi_2 n_{0,2}\} + \phi_2(\pi\phi_1 n_{0,1} + \phi_2 n_{0,2}) \\ (1-\pi+\rho\pi)\phi_1 \mathrm{E}(y_{1,2}) + \rho\phi_2 \mathrm{E}(y_{2,2}) \\ \pi\phi_1 \mathrm{E}(y_{1,2}) + \phi_2 \mathrm{E}(y_{2,2}) \\ (1-\pi+\lambda\pi)\phi_1 \mathrm{E}(y_{1,3}) + \lambda\phi_2 \mathrm{E}(y_{2,3}) \\ \pi\phi_1 \mathrm{E}(y_{1,3}) + \phi_2 \mathrm{E}(y_{2,3}) \end{bmatrix}. \tag{5.14}$$

The vector of parameters is $\boldsymbol{\theta} = [\pi, \rho, \phi_1, \phi_2]$. The derivative matrix,

$$\mathbf{D} = \frac{\partial\boldsymbol{\delta}}{\partial\boldsymbol{\theta}} = \begin{bmatrix} -(1-\rho)\phi_1 n_{0,1} & \phi_1 n_{0,1} & \dots \\ \pi\phi_1 n_{0,1} + \phi_2 n_{0,2} & 0 & \dots \\ (1-\pi+\rho\pi) n_{0,1} & \pi n_{0,1} & \dots \\ \rho n_{0,2} & n_{0,2} & \dots \end{bmatrix}, \tag{5.15}$$

has full rank 4. Therefore it is theoretically possible to estimate all four parameters.

5.3.4 Example 4: BAMS Matrix Model Example

Another example with more states is the BAMS (survival, movement, age incrementation, then births) model (Sect. 2.4.2). The state equations are

$$\mathrm{E}[\mathbf{n}_t|\mathbf{n}_{t-1}] = \begin{bmatrix} \rho(1-\mu_{1\to2})\phi_1 & \rho(1-\mu_{1\to2})\phi_2 & \rho\mu_{2\to1}\phi_3 & \rho\mu_{2\to1}\phi_4 \\ (1-\mu_{1\to2})\phi_1 & (1-\mu_{1\to2})\phi_2 & \mu_{2\to1}\phi_3 & \mu_{2\to1}\phi_4 \\ \rho\mu_{1\to2}\phi_1 & \rho\mu_{1\to2}\phi_2 & \rho(1-\mu_{2\to1})\phi_3 & \rho(1-\mu_{2\to1})\phi_4 \\ \mu_{1\to2}\phi_1 & \mu_{1\to2}\phi_2 & (1-\mu_{2\to1})\phi_3 & (1-\mu_{2\to1})\phi_4 \end{bmatrix} \mathbf{n}_{t-1}. \tag{5.16}$$

We again assume that all states are observed, so that the observation process is

$$\mathrm{E}(\mathbf{y}_t) = \begin{bmatrix} 1&0&0&0 \\ 0&1&0&0 \\ 0&0&1&0 \\ 0&0&0&1 \end{bmatrix} \mathbf{n}_t.$$

As this model has $m = 4$ states, if we use option 2, we would need the exhaustive summary terms $\boldsymbol{\delta} = [\mathbf{y}_1', \dots, \mathbf{y}_7']'$. However it is not possible to calculate the rank of the derivative matrix symbolically for this exhaustive summary. Instead consider the exhaustive summary $\boldsymbol{\delta} = [\mathbf{y}_1', \mathbf{y}_2']'$. The vector of parameters is $\boldsymbol{\theta} = [\rho, \mu_{1\to2}, \mu_{2\to1}, \phi_1, \phi_2, \phi_3, \phi_4]$. The derivative matrix $\partial\boldsymbol{\delta}/\partial\boldsymbol{\theta}$ has full rank 7. Adding an extra exhaustive summary term $\mathbf{y}_3$ adds no extra parameters, therefore the derivative matrix will still be full rank. This result is a trivial application of the extension theorem (Remark 7 of Catchpole and Morgan 1997). Therefore this model is not parameter-redundant for any $T \geq 2$ years of study.

5.3.5 *Variance Components, Non-linear Terms and Reparameterization*

The error processes $\boldsymbol{\epsilon}_t$ and $\boldsymbol{\delta}_t$ are typically assumed to have mean zero with specific variances. If these variances are not fixed known constants then the variance of $\mathbf{y}_t$ needs also to be expanded, adding extra terms to the exhaustive summary. The exhaustive summary is then

$$\boldsymbol{\delta} = [\mathrm{E}(\mathbf{y}_0), \mathrm{Var}(\mathbf{y}_0), \mathrm{E}(\mathbf{y}_1), \mathrm{Var}(\mathbf{y}_1), \mathrm{E}(\mathbf{y}_2), \mathrm{Var}(\mathbf{y}_2), \ldots]' . \tag{5.17}$$

We can also extend the method to non-linear state-space models. Suppose that $\mathbf{y}_t = h(\mathbf{x}_t) + \boldsymbol{\eta}_t$ with $\mathbf{x}_t = g(\mathbf{x}_{t-1}) + \boldsymbol{\epsilon}_{t-1}$, for suitably defined functions f and g and error processes $\boldsymbol{\eta}_t$ and $\boldsymbol{\epsilon}_t$. Then an exhaustive summary is

$$\boldsymbol{\delta} = \begin{bmatrix} \mathrm{E}(\mathbf{y}_1) \\ \mathrm{E}(\mathbf{y}_2) \\ \vdots \end{bmatrix} = \begin{bmatrix} h\{g(\mathbf{n}_0)\} \\ h[g\{g(\mathbf{n}_0)\}] \\ \vdots \end{bmatrix} . \tag{5.18}$$

The algebra of the exhaustive summary term $\mathrm{E}(\mathbf{y}_t)$ becomes increasingly complex as t increases. As mentioned previously, if exhaustive summary terms are too complex, Maple can run out of memory trying to calculate the rank of the derivative matrix. In such cases the hybrid-symbolic method of Choquet and Cole (2012) or the reparameterization method of Cole et al. (2010) could be used.

Examples of non-linear models, parameters in variance components and use of the reparameterization method are given in Cole and McCrea (2012).

5.4 Model Evaluation: Model Selection

In the analysis of complex data, there may be many competing models that can be fitted. Typically, these models represent competing biological hypotheses. For example, demographic rates are often expressed as a function of different possible covariates. We often use a logit link function for expressing the demographic rates (such as survival rates) that lie in the interval [0,1] as a function of the covariates. Often there may be a large number of possible covariates (such as environmental conditions or individual attributes) that may influence the given demographic rate. Thus, identifying the important covariates that are related to the demographic rate often provides information on the underlying dynamics of the population under study. This can be very important in determining, for example, conservation policies for managing populations and/or future prediction of the system. Alternatively, the underlying age or time dependence of the demographic parameters may be unknown and of interest. For example, we may be interested in investigating whether first-year animals have different survival rates from older animals, or whether the productivity rate declines with age. Such hypotheses can all be represented as different models, or competing hypotheses. Consider the case where we are interested in whether the survival rate is different (and lower) for individuals in their first year compared with older individuals. Then we wish to compare the two models given by:

Model 1: $\phi_a = \phi$ for all ages a;
Model 2: $\phi_1 < \phi_a = \phi$ for $a = 2, \dots,$

where the subscript denotes the age dependence. In this case we can express the above model selection problem as a hypothesis test:

$$H_0 : \phi_a = \phi \text{ for all ages } a \quad \text{vs} \quad H_1 : \phi_1 < \phi_a = \phi \text{ for } a = 2, \dots \tag{5.19}$$

where H_0 is the null hypothesis and H_1 the alternative hypothesis.

We discuss in turn the most common model discrimination techniques within both classical and Bayesian frameworks.

5.4.1 Classical Methods

Within classical data analyses, the estimation of parameters of interest is generally a two-step process in the presence of model uncertainty. The first step involves choosing a model and then, given this model, the second step involves estimating the parameters, calculating their corresponding maximum likelihood estimates using a numerical optimization algorithm. Thus, the first step involves selecting the model that is judged to be "optimal". Typically, several different models are fitted to the data, and compared using some criterion—most commonly a likelihood ratio test or information criterion. We introduce these, together with score tests and the Lasso (least absolute shrinkage and selection operator). All of these methods are based on the concept of parsimony: choosing the least complex model for which the fit is "adequate". We discuss each of these model selection techniques in turn.

5.4.1.1 Likelihood Ratio Tests

The likelihood ratio test allows hypothesis testing between two competing nested models. The idea is to compare the fit of the model to the data under the null and alternative hypotheses, where the model in the null hypothesis is nested in the model specified by the alternative hypothesis. We then assess whether the improved fit of the model under the alternative hypothesis justifies the additional parameters in this model. Formally, we calculate the statistic:

$$\Lambda = -2\log\left(\frac{L(\hat{\boldsymbol{\eta}}_0|\mathbf{y}_{1:T})}{L(\hat{\boldsymbol{\eta}}_1|\mathbf{y}_{1:T})}\right), \tag{5.20}$$

where L denotes the likelihood function evaluated at the maximum likelihood estimates $\hat{\boldsymbol{\eta}}_0$ and $\hat{\boldsymbol{\eta}}_1$ of the parameters, estimated under the null and alternative hypotheses respectively, and $\mathbf{y}_{1:T}$ denote the observed data. The statistic Λ has an asymptotic chi-squared distribution with degrees of freedom equal to the difference in the number of independent parameters between the null and alternative hypotheses. For example, suppose that we consider the hypotheses in Eq. (5.19), for which the difference in the number of independent parameters is simply equal to one. Thus, we fit the two models (where we include the additional restriction that ϕ_1 is less than ϕ in the alternative hypothesis) and calculate Λ. We compare this statistic with a χ_1^2 distribution. Assuming a 5 % significance level, the corresponding χ_1^2 value is equal to 3.84. Thus if $\Lambda \leq 3.84$, the data are considered to be consistent with the null hypothesis, but if $\Lambda > 3.84$, we have evidence against the null hypothesis.

Note that the likelihood ratio test can only be used to compare nested models. In practice, when using a likelihood ratio test where there are many possible models, a search algorithm implemented. This typically involves adding (or deleting) terms in a model to see whether the fit is significantly improved (or decreased) at some given level. However, different algorithms used to add/remove parameters can lead to the identification of different models. In this case, it can be difficult to compare the different models obtained, as they will typically be non-nested. In addition, when there are many models, this approach can be time-consuming and/or impractical. In such instances, we might focus on only a small subset of models and/or search algorithms (Edwards and Havránek 1985). Finally, we note that the likelihood ratio test relies upon asymptotic results.

5.4.1.2 Score Tests

Score tests (Rao 1948) can be used to perform hypothesis testing between two competing nested models. We adopt the same notation as above, where the parameters in the null hypothesis are denoted by $\boldsymbol{\eta}_0$ and in the alternative hypothesis by $\boldsymbol{\eta}_1$. Note that these two vectors can be of different length. For notational convenience, we set

$$l(\boldsymbol{\eta}) = \log L(\boldsymbol{\eta}|\mathbf{y}_{1:T}), \tag{5.21}$$

so that $l(\boldsymbol{\eta})$ denotes the log-likelihood function evaluated at $\boldsymbol{\eta}$. We then define the score vector,

$$\mathbf{u} = \left[\frac{\partial l(\boldsymbol{\eta})}{\partial \eta_i}\right], \tag{5.22}$$

where η_i denotes the ith component of $\boldsymbol{\eta}_1$. In other words we calculate the vector of derivatives of the log-likelihood with respect to the parameters in the model given in the alternative hypothesis. We also derive the Fisher information matrix,

$$\mathbf{I} = -\mathrm{E}\left[\frac{\partial^2 l(\boldsymbol{\eta})}{\partial \eta_i \partial \eta_j}\right]. \tag{5.23}$$

Thus the Fisher information matrix is the negative of the expectation of the Hessian (i.e. matrix of second derivatives) of the log-likelihood function.

The score statistic is given by (differentiating with respect to the parameters in the alternative model)

$$S = \hat{\mathbf{u}}_0' \hat{\mathbf{I}}_0^{-1} \hat{\mathbf{u}}_0, \tag{5.24}$$

where $\hat{\mathbf{u}}_0$ and $\hat{\mathbf{I}}_0$ denote the score vector and Fisher information matrix respectively, evaluated at the maximum likelihood estimates of the parameters under the null hypothesis, $\hat{\boldsymbol{\eta}}_0$. When there are nuisance parameters, the only optimization takes place with respect to them. To obtain the score statistic we only need to calculate the maximum likelihood estimates of the parameters in the null model, since these are used in the score statistic. The alternative model is only used in the calculation of the derivatives for $\mathbf{u}$ and $\mathbf{I}$.

Under certain regularity conditions, the score statistic has an asymptotic χ^2_ν distribution, where the degrees of freedom ν are equal to the difference in the number of parameters under the alternative and null hypotheses. See Catchpole and Morgan (1996) and McCrea and Morgan (2011) for further discussion and applications to ecological models.

The appeal of score tests is that we can compare competing nested models by fitting only the simpler model to the data. Thus, the more complex model of the alternative hypothesis need only be fitted when the null hypothesis is rejected. Note that using the observed Hessian rather than the expected Hessian can result in negative test statistics (Morgan et al. 2007).

5.4.1.3 Information Criteria

A more general approach that allows the comparison of non-nested models and does not rely on asymptotic results is the use of an information criterion. The most widely used criteria are the AIC (Akaike Information Criterion) suggested by Akaike (1973) and the BIC (Bayesian Information Criterion) of Schwarz (1978). Both these criteria consider a trade-off between the complexity and the corresponding fit of the model. Let $\hat{\boldsymbol{\eta}}$ denote the maximum likelihood estimates of the parameters. The information criteria are of the general form

$$-2 \log L_m(\hat{\boldsymbol{\eta}}|\boldsymbol{y}_{1:T}) + p^*, \tag{5.25}$$

where $L_m(\cdot)$ denotes the likelihood function evaluated at the maximum likelihood estimates of the parameters corresponding to model m (and hence $-2\log L_m(\hat{\boldsymbol{\eta}}|\boldsymbol{y}_{1:T})$ is the corresponding deviance), and p^* denotes a penalty term involving the number of parameters in the model. Let $|m|$ denote the number of parameters in model m. For the AIC statistic, the penalty term, $p^* = 2|m|$; for the BIC statistic, $p^* = |m|\log n$, where n is the number of data points. See Burnham and Anderson (2002) for further information criteria, including the (adapted) AIC: (quasi-likelihood) QAIC for over-dispersed data and (corrected) AICc for small sample sizes. For the given information criterion, the model deemed optimal is the one with smallest criterion value corresponding to an adequate fit of the model to the data (and hence a relatively large likelihood value), penalized for the number of parameters used. Different criteria may lead to different conclusions and it may be difficult to decide which is the most appropriate.

In general, the absolute value of the criterion is not important; it is intended to be used as a relative measure of fit. In the case of AIC, it is standard to give the difference in AIC values between each model and the model deemed optimal by this criterion. This is typically denoted by ΔAIC, so that the model deemed optimal has a value of ΔAIC = 0. Models with values of ΔAIC ≤ 2 are generally regarded as supported by the data (Burnham and Anderson 2002).

Typically when using an information criterion as a model selection technique, each individual model must be fitted to the data and the corresponding information criterion values calculated. However, it is possible to automate this method using trans-dimensional simulated annealing (Brooks et al. 2003). This approach extends the standard simulated annealing optimisation algorithm (Kirkpatrick 1984). The underlying idea of standard simulated annealing (i.e. for fixed-dimension problem) is to construct an algorithm that searches over some parameter space, $\boldsymbol{\Theta}$ and "freezes" at the minimum value of a given function $f(\boldsymbol{\theta})$ for $\boldsymbol{\theta} \in \boldsymbol{\Theta}$. Thus, to find the maximum likelihood estimates of the parameters, simply set $f \equiv -L$ (i.e. the minus of the likelihood function). The trans-dimensional simulated annealing algorithm extends this standard algorithm by specifying the associated function $f \equiv -2\log L_m + p^*$ and searching over both model and parameter space in order to find the model with the lowest information criterion (and the associated parameter values). Thus, instead of fitting each possible model, a search algorithm is constructed over the model space itself. We note that this algorithm can be difficult to implement and typically involves substantial pilot-tuning. However, it is very powerful, permitting a much more extensive set of models to be considered than is possible when fitting each model individually (King and Brooks 2004).

5.4.1.4 Lasso

The Lasso (least absolute shrinkage and selection operator) is a shrinkage and selection method for linear regression models, proposed by Tibshirani (1996). In linear regression, estimates of the parameters for a given model are typically obtained by minimizing the error sum of squares. For example, suppose that we have responses y_t regressed on covariates $r_{1,t}, \ldots, r_{n,t}$, such that

$$y_t = \beta_0 + \sum_{i=1}^{n} \beta_i r_{i,t} + \epsilon_t. \tag{5.26}$$

Parameter estimates for $\beta_0, \ldots, \beta_n$ are taken to be the values that minimize the sum of squares of errors:

$$\sum_{i=1}^{n} \epsilon_t^2 = \sum_{i=1}^{n} \left(y_t - \beta_0 - \sum_{i=1}^{n} \beta_i r_{i,t} \right)^2 . \tag{5.27}$$

To perform model selection using the Lasso, we ensure that covariates $r_{i,t}$ are normalized to each have mean zero and variance one, $i = 1, \ldots, n$. Then the above sum of squares is again minimized, but with the constraint that $|\sum_{i=1}^{n} \beta_i| \leq s$, where $s \geq 0$ is a tuning parameter. For large enough values of s, the constraint has no effect on the parameter estimates; however, for smaller values of s, the parameter estimates are reduced, and some of the regression coefficients will typically equal zero, thus reducing the size of the model.

In ecological applications, we typically do not have responses directly regressed on covariates, but unknown demographic parameters that are regressed on covariates. Thus, Brown (2010) has proposed a two-step process. Initially, the model is fitted assuming an arbitrary time dependence for the demographic parameters (if the covariates are time-varying), and their maximum likelihood estimates are calculated. The second step involves taking the estimated demographic parameters to be response variables (i.e. the $\boldsymbol{y}$). The standard application of the Lasso is then applied, and the regression coefficients $\beta_0, \ldots, \beta_n$ estimated as described above.

5.4.2 Bayesian Methods

The same underlying problem of model discrimination arises within the Bayesian framework. There have been a number of different approaches suggested for discriminating between competing models and algorithms proposed to implement the approaches. We discuss the two most prominent approaches. The first is an information criterion approach while the second is perhaps a more intuitive idea and a simple extension of Bayes' Theorem.

5.4.2.1 Deviance Information Criterion

Spiegelhalter et al. (2002) proposed the Deviance Information Criterion (DIC) for model selection. Its form is similar to that for other information criteria:

```
DIC = goodness of fit + penalty term for complexity.
```

To assess the goodness of fit of the model to the data, $\mathbf{y}$, the deviance is typically used. However, within the Bayesian framework, the parameters, $\boldsymbol{\eta}$, have a posterior distribution, which implies a distribution on the deviance. Thus, the goodness of fit is taken to be the posterior mean of the deviance, and often denoted `Dbar`, i.e. `Dbar` $= \mathrm{E}_{\pi}[-2\log L(\boldsymbol{\eta}|\mathbf{y})]$. The complexity of the model is represented by the effective number of parameters and denoted by p_D, given by

$$p_D = \texttt{Dbar - Dhat},$$

where `Dhat` $= -2\log L(\tilde{\boldsymbol{\eta}}|\mathbf{y})$, for $\tilde{\boldsymbol{\eta}}$ a posterior point estimate of the parameters $\boldsymbol{\eta}$ (i.e. `Dhat` is the deviance evaluated at a posterior estimate of $\boldsymbol{\eta}$). The most common choice of point estimate is the posterior mean, so that $\tilde{\boldsymbol{\eta}} = \mathbb{E}_{\pi}(\boldsymbol{\eta})$. We now take

$$\texttt{DIC} = \texttt{Dbar} + p_D = \texttt{Dhat} + 2p_D.$$

Although the DIC is appealing in its apparent simplicity, it is not without controversy. For example, the effective number of parameters is based on a single posterior point estimate of each parameter; different results can be obtained, depending on the point estimate used; and it is possible to obtain a negative value of p_D. See Celeux et al. (2006) for further discussion of the DIC, with particular reference to mixture models and models with missing data. Millar (2009) addresses the use of DIC for comparing hierarchical Bayesian models for overdispersed count data.

5.4.2.2 Posterior Model Probabilities and Bayes Factors

In the Bayesian framework, there is a natural way to deal with model uncertainty using a simple extension of Bayes' Theorem. The model can be treated as a discrete parameter and the joint posterior distribution derived over the parameters, states and model space. Letting $\boldsymbol{\eta} = (\boldsymbol{\theta}', \boldsymbol{\psi}')'$ denote the vector of parameters in the state and observation processes collectively, we note that these are now typically dependent on the model itself. Thus, we let $\boldsymbol{\eta}_m = (\boldsymbol{\theta}'_m, \boldsymbol{\psi}'_m)'$ denote the parameters in model m. Treating the model as a discrete parameter and using Bayes' Theorem, the joint posterior distribution over parameters, states and model space can be expressed as:

$$\pi(\mathbf{n}_{0:T}, \boldsymbol{\eta}_m, m|\mathbf{y}_{1:T}) \propto p(m)p(\boldsymbol{\eta}_m|m)g_0(\mathbf{n}_0|\boldsymbol{\theta}_m)\prod_{t=1}^{T} g_t(\mathbf{n}_t|\mathbf{n}_{t-1}, \boldsymbol{\theta}_m)f_t(\mathbf{y}_t|\mathbf{n}_t, \boldsymbol{\psi}_m).$$

This can be compared with the standard expression for the posterior distribution given in Eq. (4.9). The simple difference is that we now have the additional model uncertainty incorporated, considering the model as a discrete-valued parameter. Note that the priors for the parameter $\boldsymbol{\eta}$ are now conditional on the model, and we need an additional prior specified on the model component, denoted by $p(m)$.

Competing models (or hypotheses) can be quantitatively discriminated between via posterior model probabilities. Suppose that there are K possible models denoted by $m_1, \ldots, m_K$. The corresponding posterior model probability for model i is simply given by:

$$\pi(m_i|\mathbf{y}_{1:T}) \propto \int \pi(\mathbf{n}_{0:T}, \boldsymbol{\eta}_{m_i}, m_i|\mathbf{y}_{1:T}) d\mathbf{n}_{0:T} d\boldsymbol{\eta}_{m_i}. \tag{5.28}$$

In other words, the posterior model probabilities are calculated as the marginal posterior distribution over the model space. The constant of proportionality ensures that the sum over all possible models of the posterior model probabilities is equal to unity [and so is equal to the reciprocal of the sum over models of the right hand side of Eq. (5.28)].

A related discrimination measure between competing models is the Bayes factor. The Bayes factor of competing models, m_1 and m_2 say, is defined to be

$$B_{12} = \frac{\pi(m_1|\mathbf{y}_{1:T})p(m_2)}{\pi(m_2|\mathbf{y}_{1:T})p(m_1)}.$$

Thus, the Bayes factor is simply the ratio of posterior odds to prior odds for model m_1 compared to model m_2. In the case of equal prior model probabilities (i.e. no prior preference for either model), the Bayes factor reduces to the ratio of posterior model probabilities. Kass and Raftery (1995) discuss Bayes factors, and suggest that a Bayes factor of at least three is needed to provide positive posterior evidence of one model over the other.

Note that in the absence of strong prior information, a common prior to specify on the model is an equal prior probability on each model. However, this may not be the most sensible prior, depending on the model space. For example, consider the models defined by the hypotheses in Eq. (5.19), where the null model has a common survival rate over all ages, while the alternative model is that there is a distinct (lower) first-year survival rate. Suppose further that there is uncertainty as to whether the survival rates are dependent on a single covariate. Under the null model, there are two possible models (constant survival rate and covariate-dependent survival rate). However, for the alternative model, there are four possible models, depending on whether each survival rate (first-year and adult) is constant or depends on the covariate. Thus, if we place a flat prior over all models, we have an induced prior such that the alternative model is twice as likely a priori than the null model. See King and Brooks (2002) and King et al. (2006) for further discussion of similar prior specification issues.

Finally we note that posterior model probabilities (and Bayes factors) are generally sensitive to the priors specified on the parameters in the models. This is due to the integration over the parameter space in order to obtain the posterior model probabilities. As a consequence, typically models with larger numbers of parameters are automatically penalized by having a larger parameter space that is integrated over to calculate the corresponding posterior model probability. Due to the potential prior sensitivity of the posterior distribution, we would recommend that a prior sensitivity analysis is always performed when considering model uncertainty within a Bayesian framework.

5.4.2.3 Obtaining Posterior Model Probabilities: MCMC Approach

The integration needed to obtain the posterior model probabilities (and hence Bayes factors) is typically analytically intractable in most real applications. Several approaches have been developed to estimate posterior model probabilities using Monte Carlo integration. We discuss the most commonly used algorithm—reversible jump (RJ) MCMC (Green 1995), which can be seen as an extension of the Metropolis Hastings (MH) algorithm. The underlying principle of RJMCMC is identical to the standard MCMC approach: construct a Markov chain with stationary distribution equal to the posterior distribution of interest, then once the chain has converged to this distribution, take realisations of the Markov chain to be a sample from the posterior distribution. Posterior model probabilities can be estimated as the proportion of time that the constructed Markov chain is in each model.

In the presence of model uncertainty, the constructed Markov chain needs to be able to move between competing models. However, in general, the competing models will typically differ in the number of parameters and hence are of different dimensions. The MH algorithm described in Sect. 4.5.1 is only defined for moves between equal dimensions, so that it cannot be used for model moves. Thus, we use the RJ algorithm which allows us to move between the different models (of different dimensions).

Within the reversible jump algorithm, we typically divide the updating algorithm into two components. At each iteration of the Markov chain this involves:

Step 1: Updating each parameter in the current model using the MH algorithm;

and

Step 2: Updating the model using the reversible jump algorithm.

We have already seen how to update the parameters, conditional on the model, in Sect. 4.5.1. Thus, we focus on Step 2—the reversible jump step.

Suppose that at iteration i of the Markov chain (after completing step 1 above—updating each parameter in the current model), the state of the chain is $(\boldsymbol{\eta}^i, m^i)$. Given the current model, m^i, we propose a candidate model m', with probability $r(m^i, m')$ (recall that we treat the model as a discrete parameter). The associated proposed parameter values for model m' are denoted by $\boldsymbol{\eta}'$. Note that, in general, $\boldsymbol{\eta}'$ and $\boldsymbol{\eta}^i$ will contain different parameters and be of different dimension, i.e. $|\boldsymbol{\eta}'| \neq |\boldsymbol{\eta}^i|$, where $|\cdot|$ denotes the number of parameters. Thus we need to introduce additional parameters in order to "dimension match". In particular we introduce random variables (or auxiliary variables) $\mathbf{u}^i$ and $\mathbf{u}'$, with some proposal densities $q(\mathbf{u}^i)$ and $q'(\mathbf{u}')$, respectively, such that $|\mathbf{u}^i \cup \boldsymbol{\eta}^i| = |\mathbf{u}' \cup \boldsymbol{\eta}'|$. We define a bijective function g, such that $g(\boldsymbol{\eta}^i, \mathbf{u}^i) = (\boldsymbol{\eta}', \mathbf{u}')$. Thus g corresponds to the function relating how the current parameters (and auxiliary variables $\mathbf{u}^i$) map to the proposed parameters (and auxiliary variables $\mathbf{u}'$). The proposed move is accepted with probability $\alpha_m = \min(1, p_m)$, where

$$p_m = \frac{\pi(\mathbf{n}_{0:T}, \boldsymbol{\eta}', m'|\mathbf{y}_{1:T})r(m', m^i)q'(\mathbf{u}')}{\pi(\mathbf{n}_{0:T}, \boldsymbol{\eta}^i, m^i|\mathbf{y}_{1:T})r(m^i, m')q(\mathbf{u}^i)} \left| \frac{\partial g(\boldsymbol{\eta}^i, \mathbf{u}^i)}{\partial(\boldsymbol{\eta}^i, \mathbf{u}^i)} \right|.$$

The final term corresponds to the Jacobian (as a result of a change of variable argument relating to the bijective function g). If the move is accepted, we set $(\boldsymbol{\eta}^{i+1}, m^{i+1}) = (\boldsymbol{\eta}', m')$; otherwise the move is rejected and we set $(\boldsymbol{\eta}^{i+1}, m^{i+1}) = (\boldsymbol{\eta}^i, m^i)$. The RJ algorithm reduces to the MH algorithm when $|\boldsymbol{\eta}^i| = |\boldsymbol{\eta}'|$. We emphasise that this method uses only a single Markov chain irrespective of the number of possible models. The chain explores only models that are supported by the data, and so in general will not necessarily explore all possible models, particularly for large model spaces. With larger model spaces, longer simulations are typically needed to explore both the parameter and model space.

We note that, as for the MH algorithm, the proposal densities q, the model move probabilities $r(m, m')$ and the bijective function g are chosen arbitrarily. However, the choice will influence the performance of the RJMCMC algorithm. Typically, the algorithm requires some pilot tuning to ensure that the posterior distribution is properly explored. See King et al. (2009) for further discussion of implementational issues, including improving the performance of the RJ updates, examples of RJMCMC in practice for ecological examples, and other Bayesian approaches; and Fan and Sisson (2011) for a review of RJMCMC.

5.4.2.4 Obtaining Posterior Model Probabilities: SIS Approach

Posterior model probabilities can also be calculated using SIS, but using a different type of approach. Particles are simulated, not simply over the parameter range, but also from the different possible models. Recall that the model is treated as a discrete parameter. The proportion of particles that is initially simulated from each model is simply equal to the corresponding prior model probability. The standard SIS approach is then followed for all particles simulated from all models (see Sect. 4.5.5). We estimate the corresponding posterior model probabilities as the proportion of particles remaining at the end of the SIS algorithm in each given model.

We note that although the underlying idea appears to be simple, the implementation of this approach can be problematic. This is largely as a result of particle depletion, which is typically more extreme over the models compared to the parameter values. In addition, only a small number of models can be feasibly considered due to the computational expense of simulating a large number of particles from all of the possible models.

Fig. 5.1 We use data on UK lapwings to illustrate sequential importance sampling. The same data are also used in Chap. 7 to illustrate estimation of survival from mark-recovery data, and in Chap. 9 to illustrate integrated population modelling. Photo: Steve Buckland

5.4.2.5 Example

To illustrate the ideas associated with the above model discrimination techniques, we consider analyses of UK lapwing *Vanellus vanellus* (Fig. 5.1) data. These data were collected by the British Trust for Ornithology (BTO) and are described in detail by Besbeas et al. (2002) and King et al. (2008), along with the specified models for the classical and Bayesian analyses, respectively (including prior specification). The data relate to count data (modelled via a state-space model) and ring-recovery data (see Chap. 7) and an integrated modelling approach was used (see Chap. 9). We note that different state-space models (in terms of both the system process and the observation process) are used between the classical and Bayesian analyses performed, so that the results are not directly comparable. There are four demographic parameters (first-year survival probability, adult survival probability, productivity rate and recovery probability of rings from dead birds), each possibly dependent on two covariates: *year* and *fdays*. The covariate *fdays* corresponds to the number of days in a year with a mean Central England temperature below

Table 5.1 The ΔAIC values for the lapwing dataset for competing biological models in terms of the dependence of the recovery probability (κ), survival probabilities (ϕ) and productivity rate (ρ) on covariates $year$ and $fdays$

Model	ΔAIC
$\phi_1(fdays, year), \phi_a(fdays, year)/\kappa(year)/\rho$	0.00
$\phi_1(fdays, year), \phi_a(fdays, year)/\kappa(year)/\rho(year)$	0.00
$\phi_1(fdays), \phi_a(fdays, year)/\kappa(year)/\rho(year)$	0.54
$\phi_1(fdays), \phi_a(fdays, year)/\kappa(year)/\rho$	0.92
$\phi_1(fdays, year), \phi_a(fdays)/\kappa(year)/\rho(year)$	1.91
$\phi_1(fdays), \phi_a(fdays)/\kappa(year)/\rho(year)$	1.93

Table 5.2 The models with a posterior model probability of at least 4 % for competing biological models in terms of the dependence of the recovery probability (κ), survival probabilities (ϕ) and productivity rate (ρ) on covariates $year$ and $fdays$

Model	Posterior probability
$\phi_1(fdays), \phi_a(fdays, year)/\kappa(year)/\rho$	0.312
$\phi_1(fdays), \phi_a(fdays)/\kappa(year)/\rho(year)$	0.254
$\phi_1, \phi_a(fdays, year)/\kappa(fdays, year)/\rho$	0.076
$\phi_1, \phi_a(fdays)/\kappa(fdays, year)/\rho(year)$	0.066
$\phi_1, \phi_a(fdays, year)/\kappa(year)/\rho$	0.046
$\phi_1, \phi_a(fdays)/\kappa(year)/\rho(year)$	0.042

freezing, with the year starting in April (to correspond to the start of the breeding season as opposed to calendar year); and *year* is treated as a continuous covariate corresponding to a temporal trend. Note that within the Bayesian framework, we normalize the values of both of these covariates.

For the classical analyses, the models with $\Delta\text{AIC} < 2$ are given in Table 5.1. Each model is fitted to the data in turn. Allowing for each demographic parameter to be possibly dependent on each of the two covariates, there are $4^4 = 256$ possible models. Biologically implausible models are excluded, reducing the number of models to be considered.

We see that the selected models are generally all close neighbours of each other. This is often the case in such analyses.

Table 5.2 provides the posterior model probabilities obtained from the Bayesian analysis for the models with largest posterior support (see King et al. 2008). Once more we can see that the models identified are close neighbours of each other. There seems little evidence to distinguish between the two models with largest support (a Bayes factor of 1.23); however, there is posterior evidence of support for either of these models compared to all other models (Bayes factors ≥ 3.34). In addition, it is straightforward in the Bayesian approach to calculate marginal posterior probabilities of a covariate dependence, say, for each of the demographic parameters. For example, the marginal posterior probability that the first-year survival probability varies by $year$ (irrespective of whether it also depends on $fdays$ or whether other parameters in the model depend on covariates) is simply obtained by summing the posterior probabilities of the different models that have a $year$-dependent first-year survival probability.

We again note that these models are not directly comparable to those obtained in the classical analysis and given in Table 5.1 due to the different modelling assumptions that are made. However, we can see that many of the dependence structures identified in the models from both analyses are common (unsurprising in that we are largely using the same data). Interestingly, the Bayesian analysis identifies a number of models with a constant first-year survival probability. The overall (marginal) posterior probability of a constant first-year survival probability is 0.323. As a result, we also calculated the AIC of the model $\phi_1, \phi_a(fdays, year)/\kappa(fdays, year)/\rho$ within the classical framework, giving a ΔAIC of -0.08 compared to the model deemed optimal in Table 5.1 (i.e. a marginally lower AIC than the models previously fitted). However, this model was not fitted initially due to the apparent lack of biological justification—a constant first-year survival probability. Thus, this demonstrates the trade-off between fitting a number of models to the data to identify the optimal model(s) and the biological plausibility of the models. This is present in both Bayesian and classical analyses. In classical analyses, only a limited number of models is typically fitted, so that the most biologically plausible models are fitted and/or a search algorithm implemented. Using a Bayesian analysis, we can explore many more models by allowing the Markov chain to explore the set of possible models, but these should be over the set of biologically plausible models. (Models deemed not to be biologically plausible essentially have a prior model probability of 0.)

There are clearly similarities between the models identified using the classical and Bayesian analyses (even allowing for the slight differences in state-space models). However, there are also differences. This is not surprising since the AIC statistic has no direct comparison with posterior model probabilities; and the models fitted are not identical (albeit very similar).

5.5 Model Evaluation: Model Averaging

In the presence of model uncertainty, we are able to obtain parameter estimates conditional on each model. However, basing inference on only a single model deemed optimal (in some way) ignores this additional model uncertainty issue. One approach to dealing with this problem is "model-averaging", in which a parameter is estimated by taking a weighted average of the parameter estimates over all plausible models, where the weights are obtained from the model discrimination tool implemented. In particular, future predictions can be obtained that are not reliant on a single model, but are averaged over the competing models in a way that reflects the support for each of the models, and incorporates the additional model uncertainty. We discuss the different approaches for the classical and Bayesian approaches in turn.

5.5.1 *Classical Approach*

Within the classical framework, it is possible to obtain model weights when considering information criteria to discriminate between competing models. For simplicity, we focus on the AIC statistic here. Let w_i denote the weight associated with model $i = 1, \dots, K$. Buckland et al. (1997) suggested weights of the form

$$w_i = \frac{\exp(-\frac{1}{2}\ \Delta\ \text{AIC}_i)}{\sum_{k=1}^{K} \exp(-\frac{1}{2}\ \Delta\ \text{AIC}_k)}, \tag{5.29}$$

where AIC_i denotes the corresponding AIC statistic for model i. The weights are typically referred to as Akaike weights (for the AIC). These weights sum to unity, a necessary condition for associated weights. Now suppose that we are interested in obtaining an estimate of the parameter θ, which is common to all models. Let $\hat{\theta}_i$ denote the maximum likelihood estimate of the parameter in model $i = 1, \dots, K$. The model-averaged estimate of θ, denoted by $\hat{\theta}$, is given by

$$\hat{\theta} = \sum_{i=1}^{K} w_i \hat{\theta}_i. \tag{5.30}$$

In other words $\hat{\theta}$ is a weighted average of the maximum likelihood estimates of the parameter for each model, where the weights are the associated Akaike weights. Note that BIC weights can be defined in the same way. See Buckland et al. (1997) for further discussion of model averaging using information criteria.

5.5.2 *Bayesian Approach*

Recall that the parameters have a posterior distribution within the Bayesian approach. In addition, we have the corresponding posterior model probabilities for each plausible model, m_i for $i = 1, \dots, K$. We combine these to form the marginal posterior distribution of the parameter, θ, say, that is common to all models, using

$$\pi(\theta|\mathbf{y}_{1:T}) = \sum_{i=1}^{K} \pi(m_i|\mathbf{y}_{1:T})\pi(\theta|\mathbf{y}_{1:T}, m_i).$$

The term $\pi(\theta|\mathbf{y}_{1:T}, m_i)$ is simply the (marginal) posterior distribution of θ under model m_i, and $\pi(m_i|\mathbf{y}_{1:T})$ is the posterior probability for model m_i. These model-averaged distributions are easily estimated when using a RJMCMC algorithm: obtain a sample from the marginal distribution by taking each realisation of the parameter θ in the constructed Markov chain (following burn-in) irrespective of the other parameters or model state. These samples are then typically used to obtain model-averaged posterior summary statistics, such as the posterior mean and/or standard deviation.

We note that model-averaging should not be performed on any parameter without due care and attention, particularly when using posterior summary statistics to summarize the model-averaged distribution. The reasons for this are discussed next.

5.5.3 Important Comments

We emphasise that care needs to be taken when implementing a model-averaging approach. In particular, a parameter should only be model-averaged if it retains the same interpretation in all possible models. For example, suppose that we consider a covariate analysis for the first-year survival probability, where there is uncertainty in whether the survival probability is a function of winter rainfall, denoted by r_t. We propose two competing models:

$$\text{Model 1: } \phi_{1,t} = \frac{1}{1+\exp(\beta_0)};$$
$$\text{Model 2: } \phi_{1,t} = \frac{1}{1+\exp(\beta_0+\beta_1 r_t)}.$$

These are nested models, where model 1 is a special case of model 2, with $\beta_1 = 0$. It is clear that it does not make sense to obtain a model-averaged estimate of β_1 across both models. Additionally, it may not make sense to obtain a model-averaged estimate of β_0, even though this parameter is common to both models. If we normalize the covariate values for r_t, then the interpretation of β_0 can be regarded as the same in both models, in terms of the underlying (logistic) mean of the survival probability in an "average" year (i.e. when $r_t = 0$ in model 2). However, if the covariate r_t is not normalized, then the β_0 parameter in model 2 can be interpreted as a linear combination of the underlying (logistic) mean and the intercept parameter for the covariate regressor. Thus, the interpretation of the parameter is not the same across models. By contrast, obtaining a model-averaged estimate of the first-year survival probability leads to no such problems of parameter interpretation across models—the interpretation of the survival probability remains the same.

Further, we note that it will typically not be sensible to model-average parameter estimates if these are conflicting for different competing models. Consider the following (simple) illustrative example, where we have two possible models and we estimate the first-year survival probability in each. Assuming a classical analysis, we fit each model to the data and obtain their corresponding ΔAIC values. These are

Table 5.3 An illustrative example with the maximum likelihood estimate (mle) of ϕ_1 and corresponding 95 % confidence interval (CI) for two different models and corresponding ΔAIC values

Model	mle of ϕ_1 (95 % CI)	ΔAIC
1	0.45 (0.42,0.48)	0
2	0.56 (0.54, 0.58)	1

given in Table 5.3, along with the corresponding maximum likelihood estimates and 95 % confidence intervals for the first-year survival probability.

Using the standard rule of thumb regarding ΔAIC values, there is support for both models 1 and 2. However, the estimates of the survival probability are clearly very different, with non-overlapping 95 % confidence intervals. Thus, if we calculate the AIC weights, we obtain $w_1 = 0.622$ and $w_2 = 0.378$. This provides a weighted model-averaged estimate of 0.49 for the first-year survival probability, which is not supported by either model.

A similar issue arises in a Bayesian approach, where we would naturally take the posterior model-averaged mean of the given parameter. It is not the model-averaging idea that is at fault here, but the posterior point estimate that is used to summarize the posterior distribution, when the posterior distribution is bimodal (or multimodal in the more general case). See King et al. (2009) for further discussion.

5.6 Model Evaluation: Diagnostics

Using methods of Sect. 5.4, we can identify a model that is optimal according to our favoured selection technique (for example, information criterion, Lasso or score tests), or a ranking of models fitted to the data (using for example information criterion, posterior model probabilities), or a directly quantitative comparison of models (such as posterior model probabilities or weighted AIC statistic). However, these model selection techniques provide no evaluation as to whether any of the fitted models, including a model deemed optimal, actually provide a good representation of the underlying processes that generate the data—they will simply select the best model, or rank them given the defined set of possible models. All these possible models could fit the data poorly and hence be a poor representation of the underlying processes, yet we would still select a model or models and base our interpretation of the underlying processes on the corresponding results. Thus the model selection techniques of Sect. 5.4 only provide information on the *relative* fit of the model to the data. To assess whether a model describes the data well, we need an *absolute* measure of goodness of fit.

In this section we look at various absolute measures of goodness of fit. More generally, we consider various diagnostics for statistical models, where we define diagnostics as procedures for assessing the degree of consonance between what a model is assuming and inferring about the nature of reality and the available evidence from nature, e.g. measurements and observations. Hierarchical models, and state-space models in particular, present some unique issues in diagnostics when compared with simpler models such as normal linear regression models, and diagnostics in general are less well developed for hierarchical models. We begin with a brief overview of diagnostic procedures and concepts for simpler models, which will be a review for many readers, but can serve as a basis for comparing and contrasting diagnostics for state-space models in later sections.

5.6.1 Diagnostics in a Simple Linear Model Setting

Consider a simple linear model relating aerial photo estimates y of total tree volumes of Douglas fir over 1/8 hectare plots in a particular western United States forest to physical measurements of volume x (the "ground truth"), of the following form:

$$y_i|x_i \sim \text{normal}\left(\beta x_i, \sigma^2\right),$$

where observations y_i are assumed independent, and i denotes a randomly chosen plot in the forest. Thus the model assumes (1) the photo-based estimates are on average proportional to the true volumes, (2) variations around that average are normally distributed with mean 0 and variance σ^2, and (3) conditional on the true volumes, there is no between-plot correlation in the photo estimates.

Diagnostics for this model involve assessing the three assumptions:

1. Aerial estimates are proportional to true volumes on average.
2. Departures from that proportional relationship, i.e. the errors $\epsilon_i = y_i - \beta x_i$, are normally distributed with mean 0 and variance σ^2.
3. The errors ϵ_i are independent, $E[\epsilon_i|\epsilon_j] = E[\epsilon_i] = 0$ for $i \neq j$.

Typically the parameters β and σ^2 are unknown and must be estimated. For example, given a random sample of n plots, maximum likelihood estimates $\hat{\beta}$ and $\hat{\sigma}^2$ may be calculated. This is a further complication in that the errors ϵ_i are now estimated by $e_i = y_i - \hat{\beta}x_i$, called residuals. Graphical and analytical diagnostic procedures for such "linear normal models" are well established (Neter et al. 1996), and to a large degree based on examination of the residuals. For example, a plot of e_i versus x_i should be an oval-shaped cloud of points with the horizontal line at $e = 0$ being the longitudinal axis of the oval, consistent with a linear relationship between y and x and a constant variance σ^2. If the normal distribution assumption holds, e_i should be approximately normal$(0, \hat{\sigma}^2)$, which may be assessed with graphical (e.g. $q - q$ normal plots) and analytical (e.g. Kolmogorov-Smirnov goodness-of-fit test) procedures. Residuals that are large in absolute value compared with most other residuals are labelled outliers and can indicate violations of the assumptions. Outlier detection in the state-space model case is discussed in the sections on recursive residuals (Sect. 5.6.3) and Bayesian p-values (Sect. 5.6.4).

Diagnostics also include procedures that are not as obviously directed at determining whether or not model assumptions hold. Two such procedures are detection of influential observations and assessment of predictive ability. The influence of a single observation on parameter estimates, i.e. the fitted model, is also of interest. For example, if one observation in the above linear model is removed, does the estimate $\hat{\beta}$ change considerably? Cook's distance measure (Neter et al. 1996) quantifies the relative influence of individual observations.

Diagnosis of predictive ability essentially involves quantification of the degree of similarity between model-based predictions and observations. For example, in the forest volume model and writing $\hat{y}_i = \hat{\beta}x_i$, then $|y_i - \hat{y}_i|$ is a measure of prediction quality for one sampling unit. An alternative measure is to compare y_i with a prediction based on a model fit without that observation, i.e. a leave-one-out prediction; this notion is discussed further in Sect. 5.6.5. Summary, or omnibus, measures of prediction quality for an entire data set can be calculated, e.g. the sum of absolute errors or sum of squared errors. Summary measures are often calculated for a given model *relative* to a benchmark model, which is a special case of the given model. For example, a simple linear regression model is compared with a simple mean model:

$$y \sim \text{normal}\left(\beta_0 + \beta_1 x, \sigma^2\right)$$
$$y \sim \text{normal}\left(\beta_0, \sigma^2\right).$$

The well-known R^2 statistic is one such measure where the sum of squared errors of the linear regression model is compared with the sum of squared errors of the constant mean model:

$$R^2 = 1 - \frac{\sum_{i=1}^{n}\left(y_i - \left(\hat{\beta}_0 + \hat{\beta}_1 x_i\right)\right)^2}{\sum_{i=1}^{n}(y_i - \overline{y})^2}.$$

The R^2 value is bounded with $0 \leq R^2 \leq 1$. If the regression model fits the data perfectly, $R^2 = 1$, while if the fitted regression model has zero slope ($\hat{\beta}_1 = 0$) and so is identical to the constant mean model, $R^2 = 0$.

In addition to R^2, the model selection procedures discussed previously in Sect. 5.5 are to some degree procedures for assessing the *relative* predictive quality of different models. The limitations of such relative comparisons can be seen quite simply by examining four data sets constructed by Anscombe (1973). Scatterplots of the four data sets are shown in Fig. 5.2 along with the linear regression of y on x. Each data set consists of 11 (x, y) pairs and R^2 equals 0.67 in each case. Using R^2

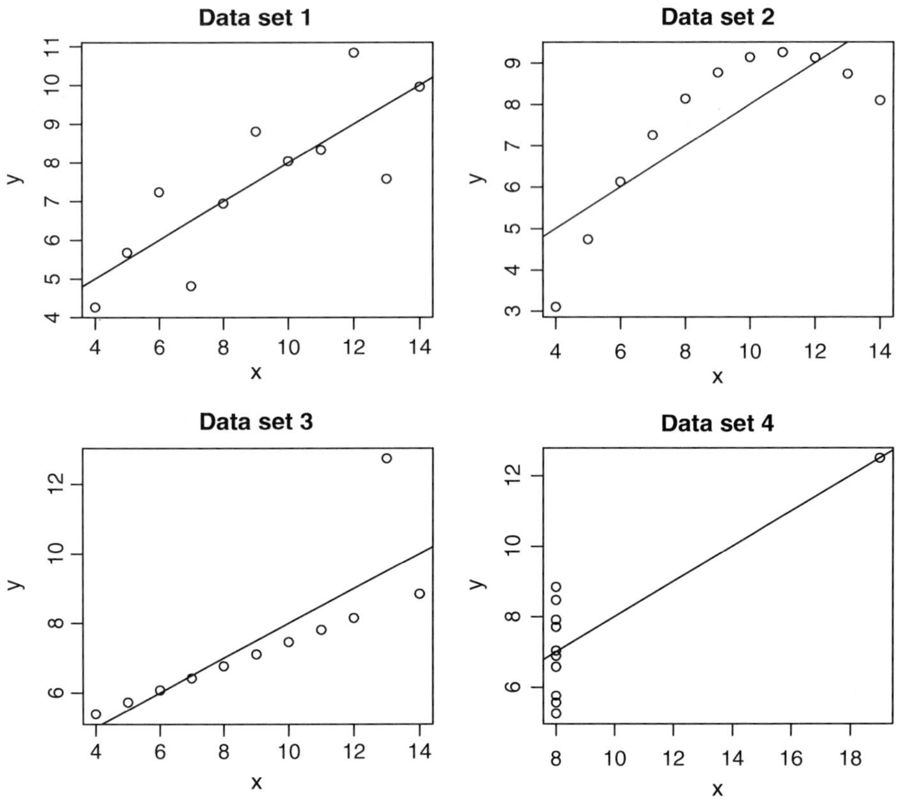

Fig. 5.2 Scatterplots of the four data sets along with the linear regression lines where in each case the R^2 value is 0.67. Taken from Anscombe (1973)

alone as a criterion would indicate that all four models are equally good. However, the plots indicate that the linear regression model is only appropriate for data set 1 and the predictive performance would be relatively low for the other three cases.

In contrast with such relative measures of goodness of fit, Sects. 5.6.3 and 5.6.4 discuss procedures for detecting individual outliers, i.e. cases where the discrepancy between a single observation and its predicted value (or predicted distribution) is large in some absolute sense. Section 5.6.5 discusses summary measures of goodness of fit that are functions of the entire data set, in a sense similar to an R^2 measure, but not constrained to [0, 1]. Before discussing such measures, we mention some unique aspects of diagnostics for SSMs.

5.6.2 Unique Aspects of SSMs

In contrast with standard linear models, state-space models pose two unique challenges for diagnostics. One challenge is that the latent or hidden variables cannot be compared to model predictions of those variables. Instead, only comparisons of the observations with their corresponding predictions can be made. Consider the following SSM with scalar-valued state and observation variables.

$$n_t | n_{t-1} \sim \text{Poisson}\,(\lambda n_{t-1})$$
$$y_t | n_t \sim \text{binomial}\,(n_t, p)\,,$$

for $t = 1, \ldots, T$, where n_0 is viewed as an unknown constant. The state process is then a stochastic exponential growth or decline model, depending on the value of λ, and p can be viewed as a detection probability. Given estimates of (λ, p, n_0), stochastic predictions of both state and observation variables can be made in a recursive manner by forward simulation: $\hat{n}_t \sim \text{Poisson}(\hat{\lambda}\hat{n}_{t-1})$ and $\hat{y}_t \sim \text{binomial}(\hat{n}_t, \hat{p})$. Suppose that to assess absolute goodness of fit, the absolute difference between the predicted and observed values was used, $e_t = |y_t - \hat{y}_t|$. Large e_t could arise for several reasons: the parameter estimates $(\hat{\lambda}, \hat{p}, \hat{n}_0)$ might be imprecise; the state process model structure might be flawed (e.g. the population multiplier should be time varying, represented by λ_t), or the observation model might be incorrect (e.g. a zero-inflated binomial distribution might be more appropriate). The inability to measure discrepancies between predicted state values and the true state values thus makes it more difficult to determine reasons for large e_t.

A second challenge, common to time series models in general, is that prediction errors propagate with time. That propagation is perhaps more easily seen by assuming a deterministic model

$$n_t = \lambda n_{t-1}$$
$$y_t = p n_t.$$

Suppose that the values of n_0 and p are known but that the estimated value of λ is biased low, $\hat{\lambda} = 0.8\lambda$. Then $\hat{y}_t = p\ 0.8^t\ \lambda^t\ n_0 = 0.8^t\ y_t$. The absolute error, $|\hat{y}_t - y_t|$ grows geometrically; e.g., the error for $t = 1$ is $0.2|y_t|$ and for $t = 10$ is $0.9|y_t|$. Recognition of this problem leads to the notion of making predictions of observations at time t conditional on the data available up to, but not including, time t, $y_{1:t-1}$. Such predictions can still suffer from imprecise parameter estimates and model mis-specification, but error propagation is reduced. The distinction can be seen as the difference between using the conditional distribution $y_t|\lambda, p, n_0$ and the conditional distribution $y_t|\lambda, p, n_0, y_{1:t-1}$. Section 5.6.3 discusses measures which use previous observations to advantage.

5.6.3 Innovations and P Scores

An *innovation* is a residual measure for SSMs that measures the discrepancy between the observation y_t and a prediction of y_t based on observations up to time $t-1$, $\hat{y}_t|y_{1:t-1}$, i.e. a filtered estimate (Engle and Watson 1981; Harvey 1989):

$$\zeta_t = y_t - \hat{y}_t|y_{1:t-1},$$

The innovation is sometimes called a one-step ahead forecast error (Durbin and Koopman 2012). If the observations are vectors instead of scalars, an innovation vector can be calculated.

A standardization of a scalar innovation yields a measure similar to a z-statistic in traditional statistics:

$$z_t = \frac{\zeta_t}{\sqrt{Var(\hat{y}_t|y_{1:t-1})}},$$

which is sometimes called a *recursive residual* (Frühwirth-Schnatter 1996). For a NDLM, z_t follows a normal(0, 1) distribution if the model is correct. For non-normal SSMs, however, the probability distribution of z_t will not be standard normal.

An alternative measure of discrepancy between an observation y_t and the model-based predictive distribution is the *P-score* (Frühwirth-Schnatter 1996):

$$u_t = \Pr(Y_t \leq y_t|y_{1:t-1}).$$

A subsequent calculation is the *transformed P-score*:

$$v_t = \Phi^{-1}(u_t),$$

where Φ is the cdf of the standard normal distribution. Extreme or outlying values are those with P-scores near 0 (unexpectedly small) or 1 (unexpectedly large). If the model is correct, and if Y_t is a continuous random variable, then $u_t \sim$ uniform(0,1) and $v_t \sim$ normal(0, 1) (Frühwirth-Schnatter 1996). The technical challenge is to determine the distribution of u_t for arbitrary SSMs. Frühwirth-Schnatter (1996) presented an approximate solution to that problem for the special case of a univariate dynamic generalized linear model (DGLM), which is a SSM with a linear normal state process model and a relatively *general* observation model (here the observation is a scalar):

$$\mathbf{n}_t \sim \text{normal}(\mathbf{A}_t\mathbf{n}_{t-1}, \mathbf{Q}_t)$$
$$y_t \sim \text{P}(\mathbf{B}_t\mathbf{n}_t),$$

where P denotes some distribution, and the observation is linearly related to the state vector. Frühwirth-Schnatter (1996) proposed standard diagnostic tools based on the P-scores and transformed P-scores, some of which we list below.

- Plot v_t against t. Outliers, autocorrelation or heterogeneity might be detected.
- Make a normal plot of v_t, i.e. plot ordered v_t against normal order statistics. Departures from a straight line might identify the same problems as the v_t versus t plot.
- Calculate a bias index, $B_n = \sqrt{n}m_1$, where $m_1 = \frac{1}{n}\sum_{t=1}^{n} v_t$, which is standard normal if the model is correct.
- Calculate a dispersion index, $D_n = \frac{nm_2 - n + 1}{\sqrt{2(n-1)}}$, where $m_2 = \frac{1}{n}\sum_{t=1}^{n}(v_t - m_1)^2$, which is asymptotically normal if the model is correct.

The sample size n and starting point in the time series in the first and second moment calculations can vary in certain situations; see Frühwirth-Schnatter (1996) for details. When the bias and dispersion indices are outside [−2, 2], that might be considered evidence for model misfit.

5.6.4 Bayesian p-Values

Some of the diagnostics in Sect. 5.6.3 are helpful for identifying outliers as well as violations of distributional assumptions such as constant variance or lack of correlation between error terms. In this section and Sect. 5.6.5 we consider what could be viewed as somewhat more global measures of model misfit, i.e. measures which are functions of the entire observation time series or at least subsequences of the time series.

Bayesian p-values were proposed by Gelman and Meng (1996) for assessing goodness of fit of models within the Bayesian framework. The approach involves simulating a series of datasets from the underlying model (and posterior distribution of the parameters) and comparing these simulated datasets with the observed dataset. Thus, this measure relies on the (posterior) predictive power of the fitted model. Let $\mathbf{y}_{1:T}$ denote the observed time series data and assume (using for example MCMC or SIS) that we can simulate from the joint posterior distribution $\pi(\mathbf{n}_{0:T}, \boldsymbol{\theta}, \boldsymbol{\psi}|\mathbf{y}_{1:T})$ to obtain a set of M samples denoted by $(\mathbf{n}_{0:T}^1, \boldsymbol{\theta}^1, \boldsymbol{\psi}^1)\ldots(\mathbf{n}_{0:T}^M, \boldsymbol{\theta}^M, \boldsymbol{\psi}^M)$. For sample $i = 1, \ldots, M$, we simulate a dataset from the given model, denoted by $\mathbf{x}_{1:T}^i$ such that

$$f(\mathbf{x}_{1:T}^i|\mathbf{n}_{0:T}^i, \boldsymbol{\psi}^i) = \prod_{t=1}^{T} f_t(\mathbf{x}_t^i|\mathbf{n}_t^i, \boldsymbol{\psi}^i),$$

dropping the dependence on the state parameters, since the observed data depend only on the underlying states and observation process parameters. If the model adequately describes the underlying processes, we would expect these simulated datasets to resemble the observed dataset. However, we need to define what is meant by “resemble” in a statistical sense, as the data are generally high-dimensional. In order to do this, we define a univariate discrepancy statistic $D(\cdot|\boldsymbol{\theta}^i, \boldsymbol{\psi}^i, \mathbf{n}_{0:T}^i)$

which is a function of the data. Note that we drop the conditioning on the state parameters $\boldsymbol{\theta}$, and write $D(\cdot|\boldsymbol{\psi}^i, \mathbf{n}^i_{0:T})$ since the data are conditionally independent of $\boldsymbol{\theta}$ given the state vector $\mathbf{n}_{0:T}$. This discrepancy statistic is evaluated at both the observed and simulated data for a given set of sampled parameter values from the posterior distribution. The discrepancy function is defined such that smaller values indicate a level of consistency between the model and the data, and larger values indicate some discrepancy between model and data. We then calculate $D(\mathbf{x}^i_{1:T}|\boldsymbol{\psi}^i, \mathbf{n}^i_{0:T}) - D(\mathbf{y}_{1:T}|\boldsymbol{\psi}^i, \mathbf{n}^i_{0:T})$ for $i = 1, \ldots, M$, i.e. the difference of the discrepancy functions evaluated at the simulated and observed data, for the given parameter values from the posterior distribution. Finally, the Bayesian p-value is given by

$$\frac{1}{M}\sum_{i=1}^{M} I([D(\mathbf{x}^i_{1:T}|\boldsymbol{\psi}^i, \mathbf{n}^i_{0:T}) - D(\mathbf{y}_{1:T}|\boldsymbol{\psi}^i, \mathbf{n}^i_{0:T})] > 0),$$

where I denotes the indicator function. In other words, the Bayesian p-value is the proportion of times that the discrepancy function for the simulated data is greater than the discrepancy function for the observed data, evaluated at the parameter values simulated from the posterior distribution; that is, the proportion of times that the simulated data are more "extreme" than the observed data, with respect to the discrepancy function D. A Bayesian p-value in the tails of the distribution (typically in the lower or upper 5 % of the distribution) suggests that the model is not a good fit to the data and the observed data are not consistent with the underlying model. Gelman et al. (2003: Chap. 6) give further discussion of Bayesian p-values.

Given a discrepancy function D, Bayesian p-values are easily calculated. Two obvious choices of discrepancy function when using state-space models are:

1. $D(\mathbf{x}|\boldsymbol{\psi}, \mathbf{n}_{0:T}) = f(\mathbf{x}|\boldsymbol{\psi}, \mathbf{n}_{0:T})$—i.e. the observation process likelihood (note that the state process likelihood is identical for the observed and simulated data, as the data are not contained within the state-process).
2. $D(\mathbf{x}|\boldsymbol{\psi}, \mathbf{n}_{0:T}) = \sum_{t=1}^{T} f(\mathbf{x}_t - \mathbf{E}_t)$ where $\mathbf{E}_t$ denotes the expected value of the observed data given the parameter values and f denotes some function. For example (assuming that the observed data are univariate over time for notational simplicity) common functions of f include the chi-squared statistic $\left(f(x_t, E_t) = \frac{(x_t - E_t)^2}{E_t}\right)$ or Freeman-Tukey statistic $\left(f(x_t, E_t) = (\sqrt{x_t} - \sqrt{E_t})^2\right)$. The expected value $\mathbf{E}_t$ is a function of the sampled posterior parameter values. Note that if we assume a normal observation error model (i.e. $\mathbf{y}_t|\mathbf{n}_t, \boldsymbol{\psi} \sim \text{normal}(\mathbf{n}_t, \sigma_y^2)$), then $\mathbf{E}_t = \mathbf{n}_t$.

It can be useful to use a range of discrepancy functions in order to assess the sensitivity of the Bayesian p-value to the discrepancy function itself.

5.6.5 Cross-Validation

There are numerous forms of cross-validation (Hastie et al. 2009) but the common principle is to quantify how well a model can predict observations that were *not* used to fit the model. The general idea is that, given a sample of n observations, we use a subset of $n_T < n$ observations to fit the model, where those n_T observations are called the *training* dataset. The fitted model is then used to make predictions of the remaining $n - n_T = n_V$ observations, the *validation* dataset. One version of cross-validation, called leave-one-out (LOO), uses all but one observation as the training set, $n_T = n - 1$, and the validation set consists of the single omitted observation. The process is repeated, one at a time for all n observations. Given a set of predictions, some measure of the discrepancy between predicted and true values is then calculated, such as square root of mean squared error (RMSE) or mean absolute prediction error (MAPE). If the discrepancy measure is considered small, the model is predicting well; conversely large discrepancy measures indicate poor predictive ability. Cross-validation applied to several models can also be used as a method for model selection (Sect. 5.4; Arlot and Celisse 2010). Here we consider cross-validation in the context of diagnosing potential problems for a single model, in particular a time series or state-space model.

de Jong (1988) developed an efficient procedure for LOO cross-validation for a linear SSM with additive mean zero errors. Due to temporal dependency, however, LOO cross-validation procedures for time series data in general will not remove the influence of the single omitted observation, say y_t, as, for example, y_{t+1} contains information about y_t. A simple solution is to remove both y_t and y_{t+1} from the training dataset and just predict y_t alone in the validation dataset. Another way to control for the temporal dependency is to use the first n_T observations as the training dataset, $y_{1:n_T}$, and the subsequent $n - n_T$ observations, $y_{n_T+1:n}$, as the validation dataset. This approach, however, will lead to propagation of errors (see Sect. 5.6.2). Hart (1994) proposed a method called *Time Series Cross Validation* (TSCV) which both avoids propagation of errors and removes the dependency of future observations, and the method is, in essence, the calculation of the best one-step-ahead predictor based on the previous data, $\hat{y}_t|y_{1:t-1}$. Cross-validation measures of discrepancy could then be the RMSE or MAPE of recursive residuals (Sect. 5.6.3). Arlot and Celisse (2010) refer to a method of cross-validation where two blocks of the time series are selected as training and validation subsets such that the degree of temporal dependence between the blocks is considered sufficiently weak. For example, let the training subset be observations $y_{m:n}$ and the validation subset be $y_{r:s}$ where $m : n$ and $r : s$ are disjoint sets and $correlation(y_i, y_j) < \epsilon$, for any $i \in m : n$ and $j \in r : s$. This method is called modified cross-validation. Hyndman (http://robjhyndman.com/researchtips/tscvexample/, accessed 31 January 2012) provides R code implementing some of the above forms of time series cross-validation.

Chapter 6
Modelling Population Dynamics Using Closed-Population Abundance Estimates

6.1 Introduction

Borchers et al. (2002) explored methods for estimating the abundance of closed populations. When that book was being written, a biologist's puzzled reaction was "What is a closed population?". In reality, there is no such thing—populations only appear closed if you look at them for a short enough time period. Look at them for longer and they will change: animals will be born, age, grow, die and move. Despite this, there are many publications devoted to closed population abundance estimation methods. This is partly because some questions of interest concern the state of a population at a single point in time, but also because it is often easier to deal with open populations by thinking of them as a series of closed populations (closed while you survey them) linked together over time by some dynamic processes. With this approach, inference about the open population proceeds in two distinct steps:

Step 1 Estimate abundance at each of a number of points in time.
Step 2 Use the series of abundance estimates to draw inferences about the population trajectory through time, and/or about the dynamic processes governing this trajectory.

Metaphors for these steps are that of taking single photographs, "snapshots", of the population abundance over time (Step 1), and then linking these separate photographs to make a movie (Step 2).

Step 2 might be purely empirical; that is, it might not involve a model of the dynamic process with any pretensions of biological relevance. Linear or log-linear regression methods are among the simplest of such models. By contrast, it could involve a model for the biological processes governing how the population changes over time; this biological processes model could include explicit sub-models for each sub-process (birth, survival, aging, etc.) or could model the net effect of these processes—as is the case with an exponential population growth model, for example. It might also be deterministic (not modelling the randomness in the population dynamics process) or stochastic.

K.B. Newman et al., *Modelling Population Dynamics: Model Formulation, Fitting and Assessment using State-Space Methods*, Methods in Statistical Ecology, DOI 10.1007/978-1-4939-0977-3_6,

6.1.1 Linking Closed-Population Methods in a State-Space Model Framework

In this chapter, we consider open population estimation in two steps, as outlined above, and we consider inferences about the population based on empirical models, deterministic population dynamics models and stochastic population dynamics models (state-space models). Here we link these two steps to our general state-space model framework and give examples of all three types of Step 2 inferential approach.

We reproduce Eqs. (3.3)–(3.5) for convenience, but with $\hat{\mathbf{y}}_t$ in place of $\mathbf{y}_t$, for reasons that will become apparent:

$$\begin{aligned} \text{Initial state pdf} : & \; g_0(\mathbf{n}_0|\theta) \\ \text{State } t \text{ pdf} : & \; g_t(\mathbf{n}_t|\mathbf{n}_{t-1}, \theta) \\ \text{Observation } t \text{ pdf} : & \; f_t(\hat{\mathbf{y}}_t|\mathbf{n}_t, \psi) \end{aligned} \tag{6.1}$$

Step 1 corresponds to the observation pdf, e.g. in a scalar setting where n_t is the true population abundance and $\hat{y}_t$ is a point estimate of n_t. Section 6.1.2 provides an overview of general methods for estimating the pdf of $\hat{y}_t$ and Sect. 6.2 provides more detailed examples, including specific probability distributions.

Step 2 corresponds to the state pdf (and often the initial state pdf). A commonly-used empirical model, for a scalar state, is one where n_t is purely a function of time, i.e. $g_t(\mathbf{n}_t|t, \theta)$. An example is a linear *trend* model:

$$n_t = \beta_0 + \beta_1 t + \epsilon_t$$

where ϵ_t is a mean zero random variable; e.g. normal$(0, \sigma_\epsilon^2)$. Simple extensions are polynomial functions of time, e.g. quadratic. Alternatively, a lognormal pdf avoids the possibility of negative predicted values:

$$n_t = \exp(\beta_0 + \beta_1 t + \epsilon_t)$$

where $\epsilon_t \sim$ normal$(0, \sigma_\epsilon^2)$. This is a specific example of a log-linear model. In general, modelling abundances as a function of time is uninformative about population processes as abundances typically are not a simple function of time. One exception is a deterministic exponential growth model $n_t = \lambda n_{t-1}$, which can be rewritten as $n_t = \lambda^t n_0$.

A slightly more informative empirical model includes environmental covariates:

$$n_t = \beta_0 + \beta_1 x_t + \epsilon_t$$

where x_t might for example be a measure of precipitation or temperature. Including such covariates, however, still fails to provide biological realism, as abundances yesterday clearly influence abundances today through the processes of mortality and reproduction: inherently n_t is a function of n_{t-1}, and the two models above, which we will call $g_t(n_t|t,\theta)$ and $g_t(n_t|x_t,\theta)$, are not true population dynamics models.

True population dynamics models explicitly state that $\mathbf{n}_t$ is a function of $\mathbf{n}_{t-1}$. As shown above, exponential growth models are perhaps the simplest. A stochastic version of a scalar case is:

$$n_t = \lambda n_{t-1} \exp(\epsilon_t),$$

which can be conveniently transformed into a linear model by a natural log transformation:

$$\ln(n_t) = \ln(\lambda) + \ln(n_{t-1}) + \epsilon_t \ .$$

Least squares (conditioning on n_{t-1}) can easily be used to estimate $\ln(\lambda)$. A popular and simple extension is the Gompertz model, which allows for density dependence (Knape et al. 2013):

$$n_t = n_{t-1} \exp\left(a + b \ln(n_{t-1}) + \epsilon_t\right) \ .$$

This simplifies by natural log transformation:

$$\ln(n_t) = a + (1+b)\ln(n_{t-1}) + \epsilon_t \quad \equiv \beta_0 + \beta_1 \ln(n_{t-1}) + \epsilon_t \ ,$$

where $\beta_0 = a$ and $\beta_1 = 1 + b$. When $b = 0$, the simple exponential model results. Additional realism is added by including covariates and explicitly including survival and birth processes (and other sub-processes, see Chap. 2). For example, in the simple exponential growth model, we might allow the population growth parameter λ to vary with time, $n_t = \lambda_t n_{t-1} \exp(\epsilon_t)$, where λ_t might be a function of environmental covariates, and/or λ_t might be written as a product of a survival probability and a birth rate, each of which is year-specific. Many of the matrix models of Chap. 2 are essentially multivariate extensions of this approach. Sections 6.3.1 and 6.3.2 include further discussion of these "Step 2" procedures.

6.1.2 A Brief Overview of Closed-Population Methods

Closed-population methods are covered in detail by Borchers et al. (2002). There are really only four kinds of closed-population abundance estimation methods that are widely used, although each has many variants. These four methods are as follows:

Plot sampling methods. These come in various guises, including quadrat sampling, circular plot sampling and strip sampling. Their essential feature is that

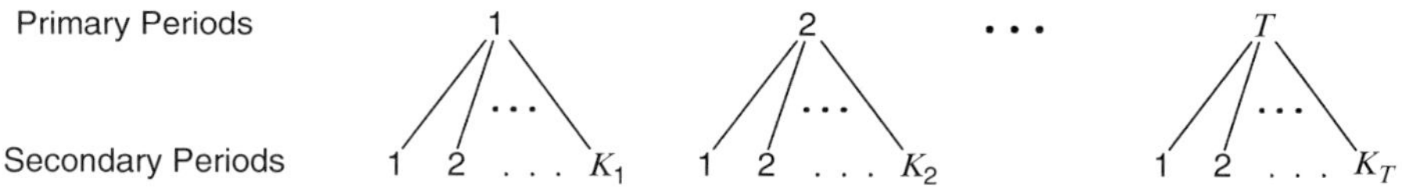

Fig. 6.1 Pollock's robust design. The population is assumed closed within primary periods but open between them. Secondary periods correspond to closed-population surveys—with multiple occasions ($K_t > 1$) for removal or mark-recapture surveys and a single occasion ($K_t = 1$) for plot sampling and distance sampling surveys

all animals within the "covered region" (that part of the survey region that is searched) are detected and counted (by assumption). All other methods involve uncertain detection within the covered region.

Distance sampling methods. Again there are various kinds, including line transect sampling, point transect sampling and cue-counting methods. Their essential feature is that, although not all animals within the covered region are detected, detection probability depends on distance of the animals from a line or point and this can be estimated from the distribution of observed distances.

Removal or harvest methods. The essential feature of these methods is that some animals are captured and removed from the population. The rate at which the number of captures declines after known numbers of removals allows estimation of the number of animals initially there. Removal methods include simple removal methods, catch-effort methods and change-in-ratio methods.

Mark-recapture methods. These methods use the recapture rate of marked individuals on subsequent survey occasions to draw inferences about the fraction of the population caught on any one occasion, and hence about the population size. There is a large literature on mark-recapture methods and an enormous variety of mark-recapture models and estimators.

Plot sampling methods and distance sampling methods allow abundance estimation from a single survey occasion. Removal and mark-recapture methods require at least two capture opportunities. To use closed-population removal and mark-recapture methods in Step 1 of the above two-step process, the time between occasions within each removal or mark-recapture survey must be sufficiently small that the population can be considered closed for this period and the sample can be considered to have come from a single point in time. Conceptually, this is similar to Pollock's (1982) "robust design" approach involving primary (e.g. years) and secondary (e.g. days) sampling periods (illustrated in Fig. 6.1) except that in our case, there are no observations that link individuals across primary periods, whereas in the open-population robust design case, animals marked during one primary period can be identified as such if they are detected during a later primary period.

We consider a population with state vector $\mathbf{n}_t$ comprising the number of individuals in each of m classes at time periods t, for $t = 1, \ldots, T$. Each of the above closed-population methods defines an observation model that can be used to obtain an estimate of abundance $\mathbf{n}_t$ or $N_t = \sum_{j=1}^{m} n_{j,t}$ which can then be used to fit a population trajectory. We illustrate with simple examples for each closed-population method before considering how these can be used to estimate population trajectories.

6.2 Observation Models

6.2.1 Plot Sampling

Consider a series of T plot samples of a population inhabiting both grassland (state 1) and woodland (state 2). We represent the population at time t using a state vector $\mathbf{n}_t = (n_{1,t}, n_{2,t})'$ $(t = 1\ldots, T)$. Suppose that for every plot sample in grassland, randomly-placed plots cover 5 % of the area and in woodland they cover 2 % of the area, so that the probability of detecting a population member in grassland is $p_1 = 0.05$ and the probability of detecting a population member in woodland is $p_2 = 0.02$. The expected counts in grassland and woodland ($y_{1,t}$ and $y_{2,t}$) are $n_{1,t}p_1$ and $n_{2,t}p_2$, and if the $n_{j,t}$ are sufficiently large, then the $y_{j,t}$ are approximately normally distributed and we can write the (approximate) observation model as

$$\mathbf{y}_t = \mathbf{O}_t\mathbf{n}_t + \boldsymbol{\epsilon}_t \qquad \text{or more explicitly,}$$

$$\begin{bmatrix} y_{1,t} \\ y_{2,t} \end{bmatrix} = \begin{bmatrix} p_1 & 0 \\ 0 & p_2 \end{bmatrix} \begin{bmatrix} n_{1,t} \\ n_{2,t} \end{bmatrix} + \boldsymbol{\epsilon}_t \tag{6.2}$$

with $\boldsymbol{\epsilon}_t = \begin{bmatrix} \epsilon_{1,t} \\ \epsilon_{2,t} \end{bmatrix} \sim \text{multivariate normal}\left(\begin{bmatrix} 0 \\ 0 \end{bmatrix}, \begin{bmatrix} \sigma^2_{y_{1,t}} & 0 \\ 0 & \sigma^2_{y_{2,t}} \end{bmatrix}\right)$.

For this survey design, the population size, $N_t = n_{1,t} + n_{2,t}$, can be unbiasedly estimated by $\hat{N}_t = \frac{y_{1,t}}{p_1} + \frac{y_{2,t}}{p_2}$ $(t = 1, \ldots, T)$. With sufficiently large N_t, the sampling distribution of $\hat{N}_t$ is well approximated by a normal distribution with mean N_t and variance $\sigma^2_{\hat{N}_t} = \sigma^2_{y_{1,t}}/p_1^2 + \sigma^2_{y_{2,t}}/p_2^2$. This leads us to the following model, which we could also think of as a kind of observation model even though the left-hand side is not really what was observed (it is a derived quantity, not the raw data—see Sect. 5.1):

$$\hat{N}_t = N_t + \epsilon_{\hat{N}_t} \quad t = 1, \ldots, T \tag{6.3}$$

where $\epsilon_{\hat{N}_t} \sim \text{normal}(0, \sigma^2_{\hat{N}_t})$. If N_t is not large, the (bias-corrected) lognormal distribution will be a better approximation to the sampling distribution of $\hat{N}_t$, and in this case

$$\hat{N}_t | N_t \sim \text{lognormal}(\log(N_t) - \hat{\sigma}^2_{\hat{N}_t}/2,\ \hat{\sigma}^2_{\hat{N}_t}). \tag{6.4}$$

Compare this to the observation model in Eq. (3.22). It is identical in form, except that (a) here we use our estimator $\hat{N}_t$ as the "observation" and (b) the variance $\sigma^2_{\hat{N}_t}$ has already been estimated in the process of obtaining the closed-population abundance estimate. We also allow the variance to be different at each time period, whereas in Eq. (3.22) it was assumed to be constant, but we could also consider a model in which the variance is constant across time; the key difference is that we use a closed-population abundance estimator as the "observation".

Note that there is a slight inconsistency in our notation in this chapter: we use $\mathbf{y}_t$ for observations throughout the book, but we use $\hat{\mathbf{y}}_t$ in Eq. (6.1) to represent a generic closed-population abundance estimator, emphasising the fact that this plays the role of the observation $\mathbf{y}_t$ in a state-space model. However, when referring to specific closed-population abundance estimators we use $\hat{N}_t$ or $\hat{\mathbf{N}}_t$, while retaining $\mathbf{y}_t$ for actual observations.

6.2.2 Distance Sampling

The two main methods of distance sampling are line transect sampling and point transect sampling. Line transect methods are slightly simpler conceptually so we will use a line transect survey for illustration. Key aspects of line transect methods are that detection probability is assumed to be one at zero distance and to decrease with distance from the line traversed by the observer, and animals are assumed to be distributed uniformly in the plane in the vicinity of lines. This latter assumption means that they are uniformly distributed with respect to distance from the line. Although it is not always the case, distances are often recorded only to within distance bands, which is the case we consider here, with each distance band constituting a "state".

Under the assumption that animals are uniformly distributed in the survey area, the state (distance band) that an animal finds itself in on survey occasion t is a random variable with known pdf (corresponding to the proportion of the survey area falling within each distance band). We then obtain a more parsimonious model if we treat the numbers of animals by state as hidden random variables than if we treat them as parameters to be estimated. This is what is done in line transect analysis, which involves a state model for animal locations (distance bands in our case). Line transect models typically separate the probability of being within some maximum distance w of the line within which animals can be detected from the line, and the probability of being beyond this distance. This is done to avoid making assumptions about the distribution of animals beyond w (see Borchers et al. 2002, for details). For our purposes, it is convenient not to make this distinction, and to this end we define the most distant distance band (state m) to be all the survey area farther than w from the line. If the m distance bands cover areas $a_1, \ldots, a_m$ and the survey region has area A, then the state model is

$$E(\mathbf{n}_t \mid N_t) = \begin{bmatrix} \frac{a_1}{A} \\ \vdots \\ \frac{a_m}{A} \end{bmatrix} N_t \tag{6.5}$$

with $(n_{1,t}, \ldots, n_{m,t}) \sim \text{multinomial}\left(N_t, (\frac{a_1}{A}, \ldots, \frac{a_m}{A})\right)$. By definition, the probability of detecting an animal in band m is zero. The probability of detecting an animal that is in band j, for $j < m$, is parameterised as a function of the distances spanned by band j and an unknown parameter vector $\boldsymbol{\psi}$. Writing this probability as $p_j(\boldsymbol{\psi})$, the expected counts on occasion t, conditional on $\mathbf{n}_t$, are

$$E(\mathbf{y}_t \mid \mathbf{n}_t) = \begin{bmatrix} p_1(\boldsymbol{\psi}) & 0 & \cdots & 0 & 0 \\ 0 & p_2(\boldsymbol{\psi}) & \cdots & 0 & 0 \\ \vdots & \vdots & \ddots & \vdots & \vdots \\ 0 & 0 & \cdots & p_{m-1}(\boldsymbol{\psi}) & 0 \\ 0 & 0 & \cdots & 0 & 0 \end{bmatrix} \begin{bmatrix} n_{1,t} \\ \vdots \\ n_{m,t} \end{bmatrix} \tag{6.6}$$

where $\mathbf{y}_t = (y_{1,t}, \ldots, y_{m,t})'$ is a vector of observed counts by distance band and $y_{j,t} \sim \text{binomial}(n_{j,t}, p_j(\boldsymbol{\psi}))$ for $j = 1, \ldots, (m-1)$ and $t = 1, \ldots, T$. By definition, $y_{m,t} = 0$. This is a random-effects model in which the states $(n_{1,t}, \ldots, n_{m,t})$ are the random effects. Denoting the pdf of $\mathbf{n}_t$ as $f_{n_t}(\mathbf{n}_t \mid N_t) = \text{multinomial}\left(N_t, (\frac{a_1}{A}, \ldots, \frac{a_m}{A})\right)$, and the pdf of $\mathbf{y}_t$ as $f_{y_t}(\mathbf{y}_t \mid \mathbf{n}_t, \boldsymbol{\psi}) = \prod_{j<m} \text{binomial}(n_{j,t}, p(\mathbf{x}_j; \boldsymbol{\psi}))$, we can write the *joint distribution of states and observations* concisely as $f_{n_t}(\mathbf{n}_t \mid N_t) f_{y_t}(\mathbf{y}_t \mid \mathbf{n}_t, \boldsymbol{\psi})$. Inference about parameters via maximum likelihood methods is based on the *marginal distribution of the observations*, equivalently the likelihood, thus summing out the states in the joint distribution, i.e.

$$L(N_t, \boldsymbol{\psi} \mid \mathbf{y}_t) \equiv f_{y_t|N_t}(\mathbf{y}_t \mid N_t, \boldsymbol{\psi}) = \sum_{\sum n_{j,t} = N_t} f_t(\mathbf{n}_t \mid N_t) f_t(\mathbf{y}_t \mid \mathbf{n}_t, \boldsymbol{\psi}). \tag{6.7}$$

where the subscript $\sum n_{j,t} = N_t$ indicates the sum over all $(n_{1,t}, \ldots, n_{m,t})$ that sum to N_t, subject to all $n_{j,t} \geq y_{j,t}$. This formulation for line transect data is similar to the formulation of Royle et al. (2004) for point transect data, the main difference being that they model density as the rate parameter of a nonhomogeneous Poisson process, whereas we model abundance rather than density and assume uniform spatial distribution for simplicity.

Mean detection probability can be estimated as $\hat{p} = \sum_{j=1}^{m} \frac{a_j}{A} p_j(\hat{\boldsymbol{\psi}})$, where $\hat{\psi}$ is the estimate of $\boldsymbol{\psi}$ obtained by maximising the likelihood, and N_t is estimated by $\hat{N}_t = \frac{y_t}{\hat{p}}$, with $y_t = \sum_{j=1}^{m-1} y_{t,j}$ being the total number of animals detected. It is common to assume that $\hat{N}_t$ is lognormally distributed with estimated variance $\hat{\sigma}^2_{\hat{N}_t}$:

$$\hat{N}_t | N_t \sim \text{lognormal}(\log(N_t) - \hat{\sigma}^2_{\hat{N}_t}/2,\ \hat{\sigma}^2_{\hat{N}_t}). \tag{6.8}$$

There is a variety of methods for estimating the variance $\sigma^2_{\hat{N}_t}$—see Borchers et al. (2002) for some details.

Equation (6.8) has identical form to Eq. (6.4), and again for the purposes of modelling dynamics across occasions, we can think of this as an observation model in which the "observation" is actually an estimate.

6.2.3 Removal Method

Because of their multi-occasion nature, the state models for even the simplest closed-population removal and mark-recapture methods involve states that change over the course of the survey within primary periods, but unlike open-population methods, the changes arise only as a consequence of the observation process. They happen because the survey itself changes the state of detected animals. In the case of removal methods, each occasion moves detected animals from being available on the next occasion to being unavailable, while in the case of mark-recapture methods, it moves them from an unmarked to a marked state.

In the simplest two-sample removal survey ($K_t = 2$), the state on occasion k (i.e. secondary period k—see Fig. 6.1) is the number of animals in the population at the start of the occasion (a scalar, not a vector) and $y_{1,t}$ and $y_{2,t}$ animals are detected and removed on the two occasions. In this case, the states for the two secondary periods in primary period t (see Fig. 6.1) are N_t and $N_t - y_{1,t}$. Assuming constant capture probability p on the two occasions, and defining $y_{0,t} \equiv 0$ for notational convenience, $(y_{k,t} \mid N_t, y_{k-1,t}) \sim \text{binomial}(N_t - y_{k-1,t}, p)$ for $k = 1, 2$. The likelihood function for N_t and p is just the product of these two binomials, and it yields the maximum likelihood estimate $\hat{N}_t = \frac{y_{1,t}^2}{y_{1,t}-y_{2,t}}$ (see Sect. 4.3.2.1, and Chap. 5 of Borchers et al. 2002).

There is a variety of ways to estimate the variance $\hat{\sigma}^2_{\hat{N}_t}$ of $\hat{N}_t$ (Chap. 5 of Borchers et al. 2002) and again it is common to assume that $\hat{N}_t$ is normally or lognormally distributed.

6.2.4 Mark-Recapture Method

Unlike removal methods, mark-recapture methods involve at least two states on each sampling occasion (secondary period) and hence a state vector on secondary period k of primary period t of $\mathbf{n}_t = (n_{1,k,t}, n_{2,k,t})'$, where $n_{1,k,t}$ is the number of unmarked animals in the population ($\equiv N_t$ for $k = 1$, $t = 1$) and $n_{2,k,t}$ is the number of marked animals in the population at the start of the secondary period ($\equiv 0$ for $k = 1$, $t = 1$). The simplest model involves just two occasions $K_t = 2$ and assumes that capture probability p is constant throughout. The number of unmarked and marked animals detected on occasion k of primary period t is $y_{1,k,t}$ and $y_{2,k,t}$, respectively. For notational convenience we define $r_{k,t} = \sum_{s<t} y_{1,k,t}$ (the total number of animals marked by the start of occasion k of primary period t), and $y_{1,0,t} \equiv 0$ (no marked animals at the start of each primary occasion).

In this case, $(y_{1,k,t}|N_t, r_{k,t}) \sim \text{binomial}(N_t - r_{k,t}, p)$, and $y_{2,k,t}|r_{k,t} \sim \text{binomial}(r_{k,t}, p)$ and the likelihood function for N_t and p is just the product of these two binomials. It yields the maximum likelihood estimator $\hat{N}_t = \frac{y_{1,1,t}(y_{1,2,t}+y_{2,2,t})}{y_{2,2,t}}$. There is again a variety of ways to estimate the variance $\hat{\sigma}^2_{\hat{N}_t}$ of $\hat{N}_t$ (see Chap. 6 of Borchers et al. 2002), and again it is common to assume that $\hat{N}_t$ is normally or lognormally distributed.

Open population mark-recapture methods are covered in Chaps. 7 and 8.

6.2.5 Summary

We have illustrated above how simple versions of each closed-population survey method leads to observation models in which the "observations" are a series of T closed-population estimates of abundance. These observation models expand in dimension when the abundance of animals in a variety of states is of interest, e.g. when abundance of different age groups or sexes, and the complexity of the closed-population methods increases when less simple versions of the methods are used. However for the purposes of fitting population dynamics models using closed-population methods applied at a series of times, the key outputs are a series of observation models in which the "observations" are (possibly vector-valued) estimates of abundances at the survey times, with all observation model parameters already estimated.

The fact that the parameters of the observation model have been estimated differentiates it from the observation models introduced in Chap. 3 where the parameters of the observation model are unknown model parameters, some of which appear in the population dynamics model and all of which are estimated concurrently with the population dynamics model parameters.

In the development of observation models for the closed-population surveys above, we implicitly assumed that estimates of the parameters of the observation models for each primary period are independent. However, it is not uncommon to allow some closed-population model parameters to be shared across primary periods and this induces dependence between the abundance estimates across these periods. The observation model for all periods is multivariate. In the case in which the closed-population abundance estimates are assumed to be normal and we denote the kth estimate $\hat{N}_{t,k}$, it is

$$\begin{bmatrix} \hat{N}_{t,1} \\ \vdots \\ \hat{N}_{t,K} \end{bmatrix} \sim \text{multivariate normal}\left(\begin{bmatrix} N_{t,1} \\ \vdots \\ N_{t,K} \end{bmatrix}, \Sigma_{\hat{N}} \right) \tag{6.9}$$

where $\Sigma_{\hat{N}}$ is the variance-covariance matrix of the closed-population abundance estimates (estimated by $\hat{\Sigma}_{\hat{N}}$ in the process of obtaining these estimates). A multivariate lognormal distribution applies in the case in which the $\hat{N}_{t,k}$ are assumed to be lognormally distributed.

We dealt only with maximum-likelihood inference for closed-population methods above. Closed-population inference using Bayesian methods is an alternative and it has some advantages over maximum-likelihood inference. These were discussed briefly in Sect. 4.3.2.1. If Bayesian closed-population methods are used, the observation models that result are posterior distributions for the (possibly vector-valued) estimates of abundances at the survey times. These can be treated as prior distributions for the abundances when drawing inferences about population trajectories and dynamics.

6.3 State Process Models

There are two distinct ways of estimating population trajectories from closed-population survey estimates: by *empirical smoothing*, in which the population trajectory is obtained by fitting some sufficiently flexible regression model without explicitly modelling population dynamics, and by fitting *population dynamics models*, which involve parameters that have biological interpretation. We start by outlining empirical smoothing methods in very broad-brush strokes and then move on to consider population dynamics models in a bit more detail.

6.3.1 Empirical Smoothing

Empirical models for population dynamics make little or no attempt to model the biological drivers of dynamics, they just provide easy means of smoothing through a series of abundance estimates. That said, the distinction between empirical models and those that do attempt to reflect biological process is not quite black and white. Linear and log-linear regression models are probably the simplest parametric models used to smooth through point estimates of abundance or density, and there is some biological basis for log-linear models, for example: the slope parameter is the per capita growth rate of a population with density-independent growth.

More flexible nonparametric models, of which generalized additive models (GAMs) are a common example, have much less biological interpretability.

Their slope at a point can be interpreted as growth in population size at that point but they contain no structure that explains how the slope arises from biological considerations and their interpretation in terms of biological processes is thus opaque.

Because of their flexibility, a key part of fitting nonparametric models is deciding how smooth to make them and this involves a trade-off (purely on the basis of the observed estimates) between what is considered underlying "trend" and short-term noise about the trend, and they do not come with a mechanism for ensuring that the degree of smoothness is consistent with what is biologically plausible (insufficient smoothing might result in biologically impossible growth rates, for example). While they can give us a reasonable picture of the population trajectory, it is difficult to draw inferences from empirically-smoothed population trajectories about biological parameters of interest such as survival or birth rates. Further, there is no basis for predicting the effect of management actions from such trajectories, because the processes affected by those actions are not modelled.

Fitting either parametric or nonparametric empirical models to time series of abundance estimates can be done treating the underlying abundances as fixed (not varying about some true but unknown trend curve) or as random (varying randomly about the true but unknown curve). For a much more full discussion of empirical smoothing methods for series of abundance estimates, see Thomas et al. (2004). See also Fewster et al. (2000) and Clarke et al. (2003) for examples of the use of empirical smoothing of closed-population abundance estimates to estimate population trends.

For the remainder of this chapter, we focus on models that arise from considerations of the biological drivers of changes in population abundance. We do this through two case studies.

6.3.2 Population Dynamics Models

It is useful to distinguish between two kinds of population dynamics model. There are those that model overall change in abundance using parameters that have biological interpretation but which do not model the sub-processes leading to the change and there are those that explicitly model the sub-processes. The particular form of the first kind of model that we focus on here is what is often called a "surplus production model" or "biomass dynamic model" (see Hilborn and Walters 1992, Chap. 8, for example). This pools birth or recruitment, survival and growth into a single function. These models can be constrained by biological considerations (population growth can be bounded, for example) and they do provide a basis for predicting the effect of management actions, but they are blunt instruments—because the sub-processes leading to changes in abundance are not modelled separately. Following Hilborn and Mangel (1997), we illustrate this kind of model using a logistic form:

$$N_t = N_{t-1} + rN_{t-1}\left(1 - \frac{N_{t-1}}{K}\right) \tag{6.10}$$

where r is the "intrinsic growth rate" (per capita rate of increase) and K is the carrying capacity. If there are removals from the population via a "catch" (c_{t-1} in year $t-1$), this is subtracted from the right-hand side of Eq. (6.10) if the catch occurs after population growth at time $t-1$, or from N_{t-1} if it occurs before population growth.

The second kind of population dynamics model we consider is that in which current abundances are functions of past abundances and the processes (or sub-processes) that cause the dynamics, the evolution in time of the population characteristics, are made explicit. In other words, the dynamics are explicit functions of processes such as survival, reproduction, maturation, or movement, and those sub-processes can be further modelled as functions of covariates, including environmental measures or management actions.

With both kinds of model, we can allow stochasticity in the population dynamics—by adding some random "process error" in the case of the first kind, or by explicitly modelling the stochasticity inherent in each of the sub-processes in the case of the second kind of model.

To illustrate the use of closed-population abundance estimates to estimate population trajectories and dynamics, we look at two case studies below, one modelling the population trajectory of wildebeest abundance in the Serengeti, the other the trajectory of an exploited gray whale population in the Pacific Ocean. In each case we consider modelling dynamics using both logistic growth models and matrix models.

6.4 Population Dynamics Model Examples

6.4.1 Wildebeest

This example and its associated data are taken from Chap. 8 of Hilborn and Mangel (1997). It concerns the Serengeti wildebeest population (Fig. 6.2), which grew substantially from the early 1960s. The growth is believed to be due to elimination of rinderpest, a disease introduced by domestic cattle, and to increasing dry season rainfall. Higher dry season rainfall provides more food at a time when animals are vulnerable to starvation, which enhances survival, particularly of calves.

The population suffered illegal hunting from 1977. The size of the resulting "catch" is not known but was estimated to be about 40,000 animals per year by Hilborn and Mangel (1997), so we assume that $c_{t-1} = 40,000$ from $t = 1978$ when we fit models to the abundance estimates.

6.4.1.1 Population Dynamics Models with Observation Error Only

In addition to the logistic growth model of Eq. (6.10), Hilborn and Mangel (1997) proposed the following population dynamics model for this population, in which $n_{2,t}$

Fig. 6.2 Spotted hyaenas hunting wildebeest in the Ngorongoro crater, near Serengeti National Park. Photo: Len Thomas

is adult abundance at time t. (The reason for the 2 subscript will become clear when we formulate a matrix model for the population dynamics below.) They specified the model as follows:

$$\text{Deterministic population dynamics model}: n_{2,t} = \phi_{1,t}\rho n_{2,t-1} + \phi_{2,t}n_{2,t-1} \tag{6.11}$$

$$\text{Observation model}: \hat{N}_t = n_{2,t} + \epsilon. \tag{6.12}$$

Here $\phi_{1,t}$ and $\phi_{2,t}$ are survival probabilities of calves and adults at time t, ρ is per-capita birth rate, $\hat{N}_t$ is the estimate of combined adult and calf abundance obtained from an aerial survey (using essentially plot sampling methods), and $\epsilon \sim \text{normal}(0, \hat{\sigma}^2_{\hat{N}_t})$, where $\hat{\sigma}^2_{\hat{N}_t}$ is the estimated standard error of $\hat{N}_t$ obtained from the survey. The $\hat{N}_t$ are independent across primary periods. Survival probability is density-dependent, and a function of the previous year's dry season rainfall (R_{t-1}). It is parameterized as follows: $\phi_{j,t} = 1.25\beta_{j,1}R_{t-1}/(1.25R_{t-1} + \beta_{j,2}n_{2,t}/A)$, where A is the area occupied by the wildebeest and $\beta_{j,1}$ and $\beta_{j,2}$ are unknown parameters (see Hilborn and Mangel 1997, p189, for the rationale for this parameterization). Dry season rainfall is shown in Fig. 6.3.

Noting that the state vector $\mathbf{n}_t$ is simply the scalar N_t here, the deterministic population dynamics model of Eq. (6.11) can be written in matrix form as follows:

$$\begin{aligned} N_t|N_{t-1} &= \mathbf{A}\mathbf{S}_t\mathbf{B}N_{t-1} \\ &= \begin{bmatrix} 1 & 1 \end{bmatrix} \begin{bmatrix} \phi_{1,t} & 0 \\ 0 & \phi_{2,t} \end{bmatrix} \begin{bmatrix} \rho \\ 1 \end{bmatrix} N_{t-1} \end{aligned} \tag{6.13}$$

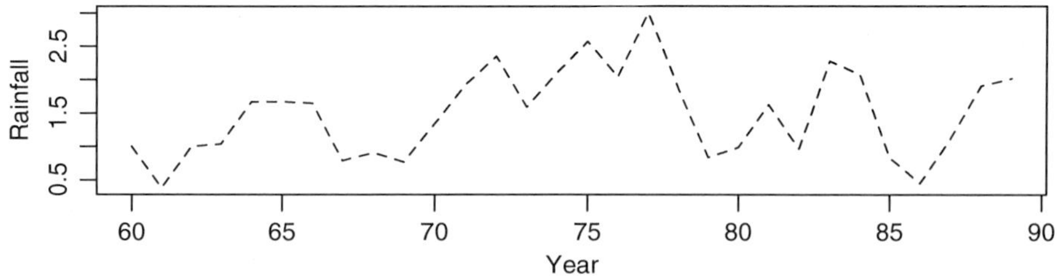

Fig. 6.3 Annual dry season rainfall in the Serengeti (100 mm)

We fitted by maximum likelihood the deterministic logistic model of Eq. (6.10), which we call Model M_L, and the deterministic matrix model above (Model M_M) with ρ fixed equal to 0.4 (following Hilborn and Mangel 1997). Because the previous year's rainfall is known to affect population growth (via survival probability), we fitted variants of the logistic model of Eq. (6.10) in which either r depends on previous year's rainfall, with $r_t = \exp(\beta_{r,0} + \beta_{r,1} R_{t-1})$ (model M_{Lr}), or both r and K depend on previous year's rainfall, with $K_t = \exp(\beta_{K,0}+\beta_{K,1} R_{t-1})$ (model M_{LrK}). We also fitted a matrix model in which only calf survival ($\phi_{1,t}$) or only adult survival ($\phi_{2,t}$) depend on the previous year's dry season rainfall (models M_{Mc} and M_{Ma}).

Among the matrix models, model Model M_{Mc} (with four estimated parameters: N_0, $\beta_{1,1}$,$\beta_{1,2}$ and $\beta_{2,1}$) is preferred on the basis of AIC, although Model M_L is preferred overall ($\Delta\text{AIC}_{Lr} = 1.9$, $\Delta\text{AIC}_{Mc} = 2.3$, and all other ΔAICs> 2.7). Despite its higher AIC, Model M_{Mc} may be substantially more useful than either of the (logistic) models with lower AIC for conservation and management purposes because it models key demographic parameters explicitly—namely per-capita birth rate (ρ), adult survival probability ($\phi_{2,t}$) and the dependence of calf survival probability ($\phi_{1,t}$) on previous year's rainfall. Because it models these parameters explicitly, it is easily extended to incorporate independent estimates of adult and calf survival probabilities that are available for some years (see Hilborn and Mangel 1997, for details); this is not the case for the logistic models.

To investigate the effect of incorporating independent survival estimates, we refitted all matrix models assuming that the survival probability estimates are lognormally distributed about the true survival probabilities. The best model among those using the survival estimates is that with only calf survival ($\phi_{1,t}$) depending on rainfall (Model M_{Mc+}). To assess the uncertainty about the estimated trajectory from this model, we resampled the model parameter values assuming normality of estimates and using the inverse Hessian matrix to estimate their variance-covariance matrix. Fits of Models M_{Lr}, M_{Mc} and M_{Mc+} are shown in Fig. 6.4. The estimated trajectory from Model M_{Mc} is more sensitive than Model M_{Lr} to the sharp changes in rainfall (shown in Fig. 6.3) from the late 1970s, although they both have the same number of parameters. It is clear from the plot that incorporating independent estimates of survival probability has a big effect on the estimated trajectory.

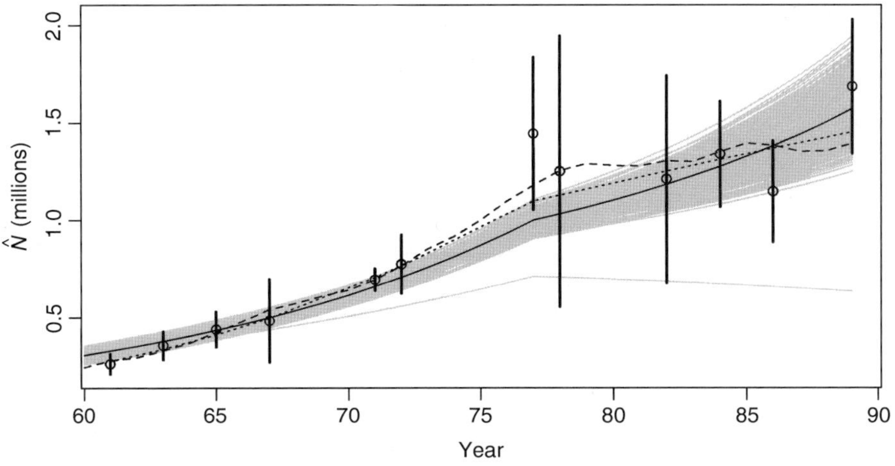

Fig. 6.4 Estimates of wildebeest abundance (*circles*) together with normal 95 % confidence intervals (*vertical lines*) and the fitted logistic growth model M_L (*dotted line*), matrix model M_{Mc} (with calf survival depending on rainfall—*dashed line*), and matrix model M_{M+} (incorporating independent estimates of survival—*solid line*). *Grey lines* are trajectories from 1,000 parametric bootstrap resamples of the parameters of model M_{M+}

To investigate the effect of incorporating independent survival estimates, we refitted all matrix models assuming that the survival probability estimates are lognormally distributed about the true survival probabilities. The only such model that did not display some problems with convergence with numerical maximization of the likelihood was a model in which neither calf nor adult survival probability depends on previous year's rainfall (Model M_{M+}). To assess the uncertainty about the estimated trajectory from this model, we resampled the model parameter values assuming normality of estimates, using the inverse Hessian matrix to estimate their variance-covariance matrix. Fits of Models M_L, M_{Mc} and M_{M+} are shown in Fig. 6.4. The estimated trajectory from Model M_{Mc} is more sensitive than that from Model M_L to the sharp changes in rainfall (shown in Fig. 6.3) from the late 1970s, although they both have the same number of parameters. It is clear from the plot that incorporating independent estimates of survival probability has a big effect on the estimated trajectory.

6.4.1.2 State-Space Model

The deterministic matrix model of Eq. (6.13) can be made into a stochastic population dynamics model by adding distributional assumptions for the birth and survival process. In the case of birth we have

$$\begin{bmatrix} E(u_{1(b)1,t}) \\ u_{1(b)2,t} \end{bmatrix} = \begin{bmatrix} \rho \\ 1 \end{bmatrix} N_{t-1} \tag{6.14}$$

and

$$\begin{pmatrix} u_{1(b)1,t} \sim \text{binomial}(n_{1,t-1}, \rho) \\ u_{1(b)2,t} = n_{2,t-1} \end{pmatrix} \tag{6.15}$$

which together define the pdf $\mathbf{g}_{1,t}(\mathbf{u}_{1,t}|\mathbf{n}_{t-1})$. In the case of survival, we have

$$\begin{bmatrix} E(u_{2(s)1,t}) \\ E(u_{2(s)2,t}) \end{bmatrix} = \begin{bmatrix} \phi_{1,t} & 0 \\ 0 & \phi_{2,t} \end{bmatrix} \begin{bmatrix} u_{1(b)1,t} \\ u_{1(b)2,t} \end{bmatrix} \tag{6.16}$$

and

$$\begin{pmatrix} u_{2(s)1,t} \sim \text{binomial}(u_{1(b)1,t}, \phi_{1,t}) \\ u_{2(s)2,t} \sim \text{binomial}(u_{1(b)2,t}, \phi_{2,t}) \end{pmatrix} \tag{6.17}$$

which together define the pdf $\mathbf{g}_{2,t}(\mathbf{u}_{2,t}|\mathbf{u}_{1,t-1})$. The final process, aging, is deterministic:

$$N_t = \begin{bmatrix} 1 & 1 \end{bmatrix} \begin{bmatrix} u_{2(s)1,t} \\ u_{2(s)2,t} \end{bmatrix} \tag{6.18}$$

and we denote this $\mathbf{g}_{3,t}(\mathbf{n}_t|\mathbf{u}_{2,t})$. The stochastic population dynamics model can be written symbolically as $\mathbf{g}_t(\mathbf{n}_t|\mathbf{n}_{t-1}) = \mathbf{g}_{3,t}(\mathbf{g}_{2,t}(\mathbf{g}_{1,t}(\mathbf{n}_{t-1})))$. Together with the observation model $\hat{N}_t|n_{2,t} \sim \text{normal}(n_{2,t}, \hat{\sigma}^2_{\hat{N}_t})$, for $t = 1, \ldots, T$ and independence between the $\hat{N}_t$, this specifies a state-space model that can be fitted using the methods described in Chap. 4.

6.4.2 Gray Whales

The National Marine Fisheries Service has conducted shore-based surveys of the stock of Eastern North Pacific gray whales (Fig. 6.5) in 23 of the 40 years between 1967/1968 and 2006/2007, from which 23 abundance estimates have been obtained. Whales are surveyed from a cliff overlooking the ocean and past which the gray whales migrate no further than about 6km offshore, making them susceptible to visual survey. This kind of survey can be viewed as a variety of line transect survey in which the animals move past the observer rather than the observer moving past the animals, and in which there is non-uniform distribution of animals with respect to distance from the observer. See Borchers et al. (2002:184–188; 240–243) for a general overview of this kind of method and Laake et al. (2013) and references therein for details of this particular survey method and analysis.

Fig. 6.5 Gray whale in Icy Bay, Alaska. Photo: NPS Photo; Scott Gende

A catch of whales is taken each year (before breeding and mortality occurs in the year, so we take catch c_t to apply to the start of year t) and we use these catch data from the 1966/1967 season onwards. To keep our model relatively simple, we consider only surveys up to and including 1997/1998, because between this and the next survey (2000/2001) there was a "catastrophic mortality event" (Punt and Wade 2010) and including this adds more complexity to the model than we would like for illustrative purposes.

6.4.2.1 Population Dynamics Models with Observation Error Only

All data used in our analysis were obtained from Punt and Wade (2010). We fitted by maximum likelihood the deterministic logistic model of Eq. (6.10) (subtracting a catch c_t at the start of year t, which is assumed to be of adults only), which we call model M_L, and an age-structured deterministic matrix model which we describe below (model M_M). It is a simplified version of the density-dependent Leslie matrix model described in Wade (2002).

We assume that sexual maturity is reached at age 6 and separately model age classes 1 through 5 and 6+ (for animals aged 6 and above). We have separate survival probabilities for ages 1 through 5 ("juveniles": ϕ_j) and 6+ ("adults": ϕ_a) and use a density-dependent birth rate, ρ_t:

$$\rho_t = \beta_0 + (\beta_1 - \beta_0)\left[1 - \frac{N_t}{K}\right] \tag{6.19}$$

where $N_t = \sum_{i=1}^{5} n_{i,t-1} + (n_{6,t-1} - c_t)$, K is the carrying capacity, β_0 is the birth rate at carrying capacity, and β_1 is the maximum birth rate. We use the model of Brandon and Wade (2006) to parameterize β_0 as a function of the survival parameters: $\beta_0 = (1 - \phi_a)/(\phi_j[1 - \phi_a^{a_{max}-5-1}])$ with $a_{max} = 200$.

This population dynamics model is a version of the BAS model introduced in Chap. 2 (with catch incorporated) and can be written in matrix form as follows.

$$\mathbf{n}_t|\mathbf{n}_{t-1} = \mathbf{B}_t\mathbf{AS}(\mathbf{n}_{t-1} - \mathbf{c}_t)$$

$$= \begin{bmatrix} 0 & 0 & 0 & 0 & \rho_t \\ 1 & 0 & 0 & 0 & 0 \\ 0 & 1 & 0 & 0 & 0 \\ 0 & 0 & 1 & 0 & 0 \\ 0 & 0 & 0 & 1 & 0 \\ 0 & 0 & 0 & 0 & 1 \end{bmatrix} \begin{bmatrix} 1 & 0 & 0 & 0 & 0 & 0 \\ 0 & 1 & 0 & 0 & 0 & 0 \\ 0 & 0 & 1 & 0 & 0 & 0 \\ 0 & 0 & 0 & 1 & 0 & 0 \\ 0 & 0 & 0 & 0 & 1 & 1 \end{bmatrix} \begin{bmatrix} \phi_j & 0 & 0 & 0 & 0 & 0 \\ 0 & \phi_j & 0 & 0 & 0 & 0 \\ 0 & 0 & \phi_j & 0 & 0 & 0 \\ 0 & 0 & 0 & \phi_j & 0 & 0 \\ 0 & 0 & 0 & 0 & \phi_j & 0 \\ 0 & 0 & 0 & 0 & 0 & \phi_a \end{bmatrix} \begin{bmatrix} n_{1,t-1} \\ n_{2,t-1} \\ n_{3,t-1} \\ n_{4,t-1} \\ n_{5,t-1} \\ n_{6,t-1} - c_{t-1} \end{bmatrix} \quad (6.20)$$

The matrix model parameters are β_0, β_1, K, ϕ_j, ϕ_a and N_0 (the population size in 1967). We assume that the population starts in a stable age structure state (obtained from the eigenvalues of $\mathbf{B}_0\mathbf{AS}$; see also Sect. 9.5.1.1) in order to distribute the initial N_0 animals across ages at the time of the first survey.

The abundance estimates are assumed to be normally distributed, initially as:

$$\hat{N}_t|N_t \sim \text{normal}(N_t,\ \hat{\sigma}^2_{\hat{N}_t}) \quad (6.21)$$

However, it is clear from the number of estimated 95 % confidence intervals for N_t from the surveys that do not intersect the fitted population trajectory, and how far these are from the line, that there is more variation in the $\hat{N}_t$ about the fitted line than can be explained by the model (see Fig. 6.6). Following Wade (2002), we add an "additional variance" parameter (σ^2_{add}) to the observation model, as follows:

$$\hat{N}_t|N_t \sim \text{normal}(N_t,\ [\hat{\sigma}^2_{\hat{N}_t} + \sigma^2_{add}]). \quad (6.22)$$

This parameter is estimated when fitting the matrix model. (Fig. 6.6 shows fitted population trajectories when σ_{add} is estimated; the trajectories when it was not in the observation model were similar.) The uncertainty associated with the estimated population trajectory was assessed by means of a parametric bootstrap of the matrix model parameters, assuming normality and using the inverse Hessian to estimate the parameter variance-covariance matrix. Trajectories from 1,000 resamples are shown as grey lines in Fig. 6.6.

Punt and Wade (2010) explain σ^2_{add} as "sources of uncertainty not captured elsewhere". It is not known whether this is a component of variance of the abundance estimates $\hat{N}_t$ that is not captured in $\hat{\sigma}^2_{\hat{N}_t}$ (i.e. $\hat{\sigma}^2_{\hat{N}_t}$ is a negatively biased

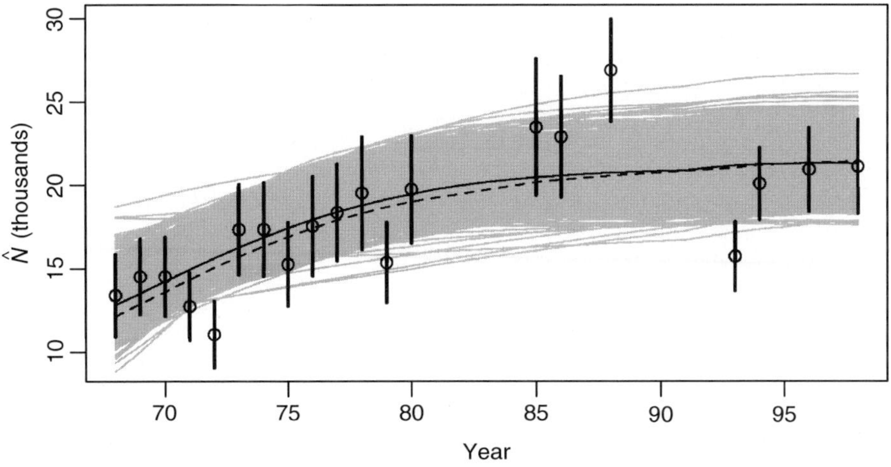

Fig. 6.6 Estimates of Eastern Pacific gray whale abundance (*circles*) together with normal 95 % confidence intervals (*vertical lines*), the fitted deterministic logistic growth model (*dashed line*), the age-structured deterministic matrix model (*solid line*), and population trajectories from 1,000 parametric bootstrap resamples of the matrix model parameters

estimator of $\sigma^2_{\hat{N}_t}$) or a failure to capture the stochasticity present in the population dynamics. The latter problem can be addressed by building a stochastic population dynamics model.

6.4.2.2 State-Space Model

A normal dynamic linear state-space model (NDLM) was fitted to the gray whale data using expected value vectors and covariance matrices for the state process that were calculated assuming binomial survival processes for each age class and a binomial birth process. A model of the form shown in Eqs. (4.11)–(4.13), which we reproduce here,

$$\begin{aligned}\mathbf{n}_0 &\sim \text{normal}\,(\boldsymbol{\mu}_0, \mathbf{Q}_0)\\ \mathbf{n}_t &\sim \text{normal}\,(\mathbf{A}_t\mathbf{n}_{t-1}, \mathbf{Q}_t)\\ \mathbf{y}_t &\sim \text{normal}\,(\mathbf{B}_t\mathbf{n}_t, \mathbf{R}_t)\end{aligned}$$

was fitted using the following expected value vector:

$$E[\mathbf{n}_t|\mathbf{n}_{t-1}] = \mathbf{A}_t\,[\mathbf{n}_{t-1} - \mathbf{c}_t] = \begin{bmatrix} 0 & 0 & 0 & 0 & \rho_t\phi_j & \rho_t\phi_a \\ \phi_j & 0 & 0 & 0 & 0 & 0 \\ 0 & \phi_j & 0 & 0 & 0 & 0 \\ 0 & 0 & \phi_j & 0 & 0 & 0 \\ 0 & 0 & 0 & \phi_j & 0 & 0 \\ 0 & 0 & 0 & 0 & \phi_j & \phi_a \end{bmatrix} \begin{bmatrix} n_{1,t-1} \\ n_{2,t-1} \\ n_{3,t-1} \\ n_{4,t-1} \\ n_{5,t-1} \\ n_{6,t-1} - c_t \end{bmatrix}$$

and covariance matrix, $\mathbf{Q}_t$,

$$\begin{bmatrix} \begin{array}{c}\rho_t\phi_j(1-\rho_t\phi_j)n_{5,t-1}+ \\ \rho_t\phi_a(1-\rho_t\phi_a)n_{6,t-1}\end{array} & 0 & 0 & 0 & 0 & \begin{array}{c}\rho_t(Var(n_{6,t})+E[n_{6,t}]^2) \\ -E[n_{1,t}]E[n_{6,t}]\end{array} \\ 0 & \phi_j(1-\phi_j)n_{1,t-1} & 0 & 0 & 0 & 0 \\ 0 & 0 & \phi_j(1-\phi_j)n_{2,t-1} & 0 & 0 & 0 \\ 0 & 0 & 0 & \phi_j(1-\phi_j)n_{3,t-1} & 0 & 0 \\ 0 & 0 & 0 & 0 & \phi_j(1-\phi_j)n_{4,t-1} & 0 \\ \begin{array}{c}\rho_t(Var(n_{6,t})+E[n_{6,t}]^2) \\ -E[n_{1,t}]E[n_{6,t}]\end{array} & 0 & 0 & 0 & 0 & \begin{array}{c}\phi_j(1-\phi_j)n_{5,t-1}+ \\ \phi_a(1-\phi_a)(n_{6,t-1}-c_t)\end{array} \end{bmatrix}$$

where $E[n_{1,t}] = \rho_t(\phi_j n_{5,t-1} + \phi_a(n_{6,t-1} - c_t))$ and $E[n_{6,t}] = \phi_j n_{5,t-1} + \phi_a(n_{6,t-1} - c_t)$. Details of these derivations can be found on the book website (Sect. 1.2).

The observation model is identical to the model used in Sect. 6.4.2.1 (the observation error only model), namely,

$$\hat{N}_t|\mathbf{n}_t \sim \text{normal}\left(\begin{bmatrix} 1 & 1 & 1 & 1 & 1 & 1 \end{bmatrix}\mathbf{n}_t, \sigma^2_{\hat{N}_t} + \sigma^2_{add}\right).$$

We note that the birth rate parameter ρ_t is a function of $\mathbf{n}_{t-1}$, i.e. the birth process is density-dependent, and the distribution of $\mathbf{n}_t$ is *conditionally* multivariate normal. This NDLM is an example of a conditionally Gaussian SSM, which is discussed further in Sect. 9.7.1.

The Kalman filter was used to calculate the negative log likelihood function (as shown in Eqs. (4.18) and (4.19)) with the initial variance matrix, Q_0, set equal to 0, and the initial population abundance vector, $\mu_0 \equiv \mathbf{n}_0$, calculated as in Sect. 6.4.2.1. Gaps in the estimates of the population abundances for several years (i.e. missing observations) were handled using the method of Harvey and Pierse (1984). Maximum likelihood estimates of the six parameters, $(\phi_j, \phi_a, \beta_1, K, N_0, \sigma_{add})$, were calculated using `R` and the code is available on the book website (Sect. 1.2). When fitting the NDLM, an estimate of ρ_t less than 0 occurred for the year 1989 when the fitted total exceeded the carrying capacity, K, and this possibility is a limitation of this particular NDLM approximation. However, optimization runs which enforced $\rho_t \geq \beta_0$ (see Eq. (6.19)), namely

Table 6.1 Maximum likelihood estimates of parameters for the observation error-only and NDLM models for the gray whale population (for the period 1968–1998)

Parameters	Observation error-only	NDLM
ϕ_j	0.9999	0.8281
ϕ_a	0.9400	0.9582
β_1	0.3418	0.9369
K	21.6391	22.2234
N_0	11.6036	13.0971
σ_{add}	2.4018	0.0360

The values for K and N_0 are in 1000s of whales

$$\rho_t = \max\left(\beta_0, \beta_0 + (\beta_1 - \beta_0)\left[1 - \frac{N_t}{K}\right]\right)$$

yielded quite similar maximum likelihood estimates, but that constraint causes a discontinuity in the likelihood function which makes calculation of the Hessian matrix impossible.

Table 6.1 compares parameter estimates for the observation error-only model and the NDLM. Figure 6.7 plots the NDLM filtered estimates of total abundance along with the observations. Estimates of carrying capacity (K) and the initial total population abundance (N_0) are relatively similar for the two approaches, while estimates of the maximum fecundity parameter (β_1) and the additional observation noise term (σ_{add}) are quite dissimilar. That the NDLM estimate of σ_{add} is two orders of magnitude smaller than for the observation error-only model can be attributed to the explicit accounting for environmental variation in the population dynamics, which has thus been separated from sampling error in the abundance estimates.

Difference in estimates of ϕ_j and β_1 may be partly due to negative correlation (in the NDLM the Pearson correlation coefficient between ϕ_j and β_1 was -0.77), indicative of some degree of parameter indeterminacy. Such indeterminacy is not surprising given that the observations are estimates of the total population, not estimates of age-specific abundances. Different combinations of survival probabilities and fecundity rates, e.g. increases in ϕ_j offset by decreases in β_1, can yield quite similar total population numbers. Quantification of such relationships between parameter estimates is useful for identifying data needs and stimulating future data collection. For example, natural markings might be exploited to conduct a mark-recapture study to estimate annual survival probabilities, ϕ_j and ϕ_a. The newly acquired data could then be combined with existing data to fit state-space models where the correlation between parameter estimates is lessened. Chapter 9 discusses such integrated analyses.

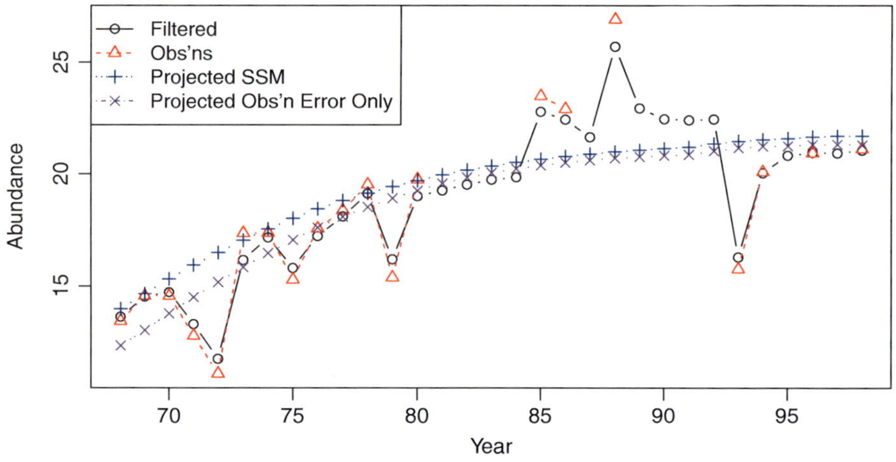

Fig. 6.7 Kalman filter estimates of annual Eastern Pacific gray whale abundances (*hollow circles*) based on an NDLM for the years 1968 through 1998, with year-specific field-based estimates (*triangles*). Projected abundances based on estimates of the initial population total N_0 and parameter estimates are contrasted for the NDLM (+'s) and the observation error-only model (x's)

6.5 Discussion

It is often convenient to draw inferences about population dynamics using closed-population abundance estimation methods, particularly when individuals cannot be linked across (primary) sampling occasions. While a single model that relates observations (rather than abundance estimates) to population dynamics parameters is intellectually appealing, breaking the inference problem into two stages and then using the "observation" model that arises from closed-population estimation can make inference substantially easier, and this is often the main motivation for adopting this strategy.

State-space modelling is slightly different when using closed-population estimators because the "observations" are really estimates in this case and all parameters of the observation model will have been estimated prior to fitting the state-space model. As Knape et al. (2013) have noted, the information loss from using estimates (derived quantities) instead of the observations (the raw data) is often minor. Fitting the population dynamics model can still be informative about parameters of the observation model. In the case of the gray whales, for example, fitting the dynamics model suggests that the variance of the closed population abundance estimates might have been underestimated. The SSM framework is potentially superior to an observation error-only model in that it allows variation in population dynamics, namely environmental and demographic stochasticity or so-called process error, to be modelled distinct from observation error. The observation error-only model necessarily combines process error and observation error. In the whale data set it was useful to have year-specific estimates of observation error (i.e. estimation error). The addition of a separate variance term (σ^2_{add}) to the observation model error was an attempt to accommodate variance over and above that estimated from surveys, although it was not clear whether this additional variance is due to underestimation of the observation error or due to neglected process error. However, the use of a constant variance term for process variation is necessarily inadequate considering that such variation in this case reflected demographic variation (in numbers surviving and numbers born), which varies as a function of abundance. A more realistic SSM for the whale data would allow for environmental variation in survival probabilities and birth rates, i.e. a hierarchical SSM might be preferable [Eq. (3.6)].

In both the examples considered here, the simpler model that does not model sub-processes (Model M_L) had lower AIC than the matrix population model (Model M_M; ΔAIC=5.5 in the case of the gray whales). Nevertheless, the latter models, which do model population sub-processes, have substantial advantages over the former. They can enhance understanding of population dynamics processes, they provide a tool for answering ecological questions about the nature of the dynamics, and can hence inform management and conservation decisions in a way that the simpler models cannot. For example management actions can be translated into effects on parameters of various sub-processes, e.g. survival probabilities and birth rates, by modelling such parameters as functions of management-related covariates. They also readily allow additional data on sub-processes to be incorporated in inference—independent estimates of survival probabilities that are available in the case of the wildebeest example.

Chapter 7
Estimating Survival Probabilities from Mark-Re-Encounter Data

There are many reasons why we are interested in how long wild animals survive. A particularly pressing one is so that we can evaluate the effects of climate and anthropomorphic changes. In an early example, North and Morgan (1979) demonstrated a link between winter temperature and the survival of grey herons, *Ardea cinerea*, using point-process models as well as logistic regression, as in Sect. 5.1. As a further example, one might be interested in calibrating the effect of a change in hunting regulations on survival probability of wild fowl. In order to study the survival of wild animals, we normally need to be able to identify individuals uniquely. This may be done by means of recognising the patterns of spots on a cheetah, the belly patterns of newts, or the DNA left in the field, for example. In many cases, animals do not exhibit recognisable individual characteristics, and so are given identifying marks. This can be done in many different ways, for instance by attaching a ring to the leg of a bird, a collar to a deer or goose, or an ear tag to a sheep. Marked in this way, individual animals may be identified and recorded when re-encountered again, alive or dead. In contrast with human demography, typically we do not know the fate of most wild animals in a population of interest, or even most of the marked or identified animals, because there may be few or no re-encounters with such animals. Marking with radio tags, combined with subsequent radio tracking, is an exception.

We describe mark re-encounter methods that are used for estimating the survival of wild animals. Probability models are formed for data obtained from observations made of uniquely identified animals, and are fitted to the data using either maximum likelihood or modern Bayesian methods. It is shown how models may be expressed in state-space form.

K.B. Newman et al., *Modelling Population Dynamics: Model Formulation, Fitting and Assessment using State-Space Methods*, Methods in Statistical Ecology,
DOI 10.1007/978-1-4939-0977-3_7,

7.1 Modelling Survival

Many wild animal populations experience a natural yearly life-cycle, and their survival is often reported as annual survival, corresponding to whether or not an animal survived a particular year of life.

7.1.1 Ring-Recovery Data

The ring-recovery data of Table 7.1 are taken from Catchpole et al. (1999). Here marked animals are only ever re-encountered dead. Thus for example, in 1964, the second year of this study, 1285 lapwings were ringed in Britain and released. Of those that died in their first year of life, the deaths of 20 were reported to the British Trust for Ornithology; of those that survived their first year of life but died in their second year, three were reported, and so on. If we assume that birds die independently of one another, then the appropriate probability distribution to describe how deaths are distributed over the years, from any one cohort of ringed birds, is the multinomial distribution. We wish to estimate probabilities of annual survival from ring-recovery data, and this involves writing the multinomial cell probabilities in terms of appropriate annual survival probabilities and recovery probabilities. A striking feature of Table 7.1 is the relatively large numbers along the leading diagonal of the table. It is often the case that there is appreciable mortality of wild animals in the first year of life. In such cases, we would want to include some form of age-dependence in the survival probabilities. For example we might have a survival probability ϕ_1 for animals in their first year of life, and an annual survival probability ϕ_a for all older animals. If we also assume a constant recovery

Table 7.1 Ring-recovery data for British lapwings *Vanellus vanellus*, uniquely ringed in the nest during the years 1963–1973

Year of	Number	(Year of recovery)										
ringing	ringed	64	65	66	67	68	69	70	71	72	73	74
1963	1147	14	4	1	2	1	0	1	1	0	0	0
1964	1285		20	3	4	0	1	1	0	0	0	0
1965	1106			10	1	2	2	0	2	2	1	1
1966	1615				9	7	4	2	1	1	0	0
1967	1618					12	1	6	2	0	0	1
1968	2120						9	6	4	0	2	2
1969	2003							10	8	5	3	1
1970	1963								8	3	2	0
1971	2463									4	1	1
1972	3092										7	2
1973	3442											15

Data provided by the British Trust for Ornithology

Table 7.2 Multinomial reporting probabilities for a model in which there is probability ϕ_1 that an animal survives its first year of life and probability ϕ_a that an animal survives any further year of life, illustrated for just one cohort of released animals and 4 years of recovery

Year	1	2	3	4	
History	D-R	L-D-R	L-L-D-R	L-L-L-D-R	Other
Probabilities	$(1-\phi_1)\kappa$	$\phi_1(1-\phi_a)\kappa$	$\phi_1\phi_a(1-\phi_a)\kappa$	$\phi_1\phi_a^2(1-\phi_a)\kappa$	$1-(1-\phi_1\phi_a^3)\kappa$

The parameter κ denotes the probability that a dead animal has its death recorded. The alphabetic labels for the history are D for Die, L for Live, and R for Recorded

probability κ, the multinomial probabilities are illustrated in Table 7.2, for the case of a single cohort of released animals and 4 years of recovery. The expression $\chi = 1 - (1 - \phi_1\phi_a^3)\kappa$ is the probability that an animal is not reported dead during the study, either because it was still alive at the end of the study or because it died but was not reported.

If there are T years of recovery, and m_t animals reported dead in the tth year of the study, then the likelihood corresponding to a single cohort of marked animals is given by

$$L(\phi_1, \phi_a, \kappa | m_1, \ldots, m_T) \propto \{\kappa(1-\phi_1)\}^{m_1} \prod_{t=2}^{T} \{\kappa\phi_1\phi_a^{t-1}(1-\phi_a)\}^{m_t} \chi^{c-m_.}, \quad (7.1)$$

where $m_. = \sum_{t=1}^{T} m_t$, and c is the cohort size. The missing constant of proportionality is the multinomial coefficient which does not involve the model parameters. For each cohort we form a separate multinomial likelihood, of the form of Eq. (7.1), and then following the additional assumption of independence between cohorts, the likelihood corresponding to the entire recovery table is given by the product of the multinomial likelihoods for each of the cohorts. This product-multinomial likelihood may then be maximized numerically to produce the maximum-likelihood estimates of the three model parameters. Alternatively it can be combined with prior distributions on the model parameters in order to form a posterior distribution in Bayesian analysis.

7.1.2 Capture-Recapture Data

The data of Table 7.3 are taken from Lebreton et al. (1992) and describe the recaptures of European dippers *Cinclus cinclus* (Fig. 7.1). These data result from a study of adult birds of unknown age which have been previously marked uniquely with rings. Here marked birds are only re-encountered alive, and after recapture they become part of the next cohort of released birds.

Table 7.3 Capture-recapture data for dippers for years 1981–1987

Year of release	Number released	Year of recapture 82	83	84	85	86	87
1981	22	11	2	0	0	0	0
1982	60		24	1	0	0	0
1983	78			34	2	0	0
1984	80				45	1	2
1985	88					51	0
1986	98						52

Fig. 7.1 Dippers are associated with clear, usually fast-flowing water, and walk underwater to catch their prey. Photo: Steve Buckland

The data may be modelled similarly to the ring-recovery data. In both cases, the model involves probabilities of annual survival, but in this case, recovery probabilities are replaced appropriately by recapture probabilities.

A display of data in this form is often called an m-array. The form of Table 7.3 is essentially the same as that of Table 7.1, and the modelling is structurally the same, involving a likelihood which is a product of multinomial likelihoods, one from each cohort. The dipper data are of adults only. Suppose we assume that they share a common annual survival probability ϕ_a, and that all birds have the same probability of recapture p in each year. Then under this simple model, and for a single cohort with 4 years of recapture data, the multinomial probabilities corresponding to first recapture in years $1, 2, 3, 4$ are respectively $\phi_a p, \phi_a^2(1-p)p, \phi_a^3(1-p)^2 p$ and $\phi_a^4(1-p)^3 p$, while those not recaptured have probability $\chi = 1 - \sum_{t=1}^{4} \phi_a^t p(1-p)^{t-1}$.

(Note that, once recaptured, a bird is released as a member of a new cohort under this model, hence it cannot be recaptured more than once in a given cohort.)

The likelihood $L(\phi_a, p|\text{data})$ can then be formed. It may be maximized numerically, to give the maximum-likelihood estimates, or be used in a Bayesian analysis in the usual way. For illustration, if we have just one cohort of size c of marked animals and there are m_t recaptures at time $t = 1, \ldots, T$, then the likelihood is given by

$$L(\phi_a, p|\text{data}) \propto \prod_{t=1}^{T} \{\phi_a^t p(1-p)^{t-1}\}^{m_t} \chi^{c-m_.}, \tag{7.2}$$

where $m_. = \sum_{t=1}^{T} m_t$, and $\chi = 1 - \sum_{t=1}^{T} \phi_a^t p(1-p)^{t-1}$.

As in the last section, the likelihood corresponding to the entire table is the product of the multinomial likelihoods that result from each cohort. When animals of unknown age are marked, then the Cormack-Jolly-Seber (CJS) model results if all of the model parameters are time-dependent (Cormack 1964; Jolly 1965; Seber 1965). A strong assumption of capture-recapture models is that animals remain faithful to the study area. If that assumption is violated, survival estimates are biased low, because we are in reality estimating the probability that an animal both survives and remains in the study area.

7.1.3 Models for Life-History Data

Re-encounters may include recapture, recovery and resighting of the same individuals. An example in which recoveries and recaptures occur together is provided in Table 7.4 which shows data on shags *Phalacrocorax aristotelis* (Fig. 7.2), arising from the Isle of May Long-Term Study of the Centre for Ecology and Hydrology in Scotland. The data take the form of individual life histories. A "1" indicates capture or recapture, a "2" indicates recovery, and a "0" indicates that the bird was not recorded in that year. This notation may be extended to include resightings, and also to cover re-encounters at more than one location. We assume here that there is no emigration of animals; for models that include migration, see e.g. Barker (1997), Burnham (1993) and Kendall et al. (1997). Suppression of 1's (2's) in the life-history data of Table 7.4 results in recovery (recapture) data alone.

We suppose that there are T sampling times, $t_1, \ldots, t_T$, with recovery information now possibly extending further to K. We define cohort c as the set of animals first marked at time $t_c, c = 1, \ldots, C$. Then for an animal from cohort c, and $j = c, \ldots, K-1$,

$\phi_{c,j}$ = Pr(an animal alive at t_j survives until t_{j+1}),
$\kappa_{c,j}$ = Pr(an animal which dies in (t_j, t_{j+1}) has its death reported),
while for $j = c+1, \ldots, K$,
$p_{c,j}$ = Pr(an animal alive at t_j is captured at t_j), and $q_{c,j} = 1 - p_{c,j}$.

Table 7.4 Illustrations of life histories for five different shags

Capture occasions						
Cohort	t_1	t_2	t_3	t_4	t_5	t_6
1	1	0	1	0	2	0
	1	1	0	0	1	1
2	0	1	2	0	0	0
	0	1	0	0	1	0
	0	1	0	0	0	0

In the first cohort, both birds are marked at time t_1, while in the second cohort, the three birds are all marked at time t_2. A "1" indicates capture, a "2" indicates recovery and a "0" indicates that the bird was not recorded in that year

Fig. 7.2 Shags are one of the species recorded in the Isle of May Long-Term Study, allowing accurate estimation of annual survival rates. Photo: Steve Buckland

Catchpole et al. (1998a) show that we can specify the likelihood as

$$L(\boldsymbol{p}, \boldsymbol{\phi}, \boldsymbol{\kappa}|data) \propto \prod_{c=1}^{C} \left[\prod_{j=c}^{K-1} \{\alpha_{c,j}(1-\phi_{c,j})\kappa_{c,j}\}^{d_{c,j}} \prod_{j=c}^{T} \{\alpha_{c,j}\chi_{c,j}\}^{v_{c,j}} \prod_{j=c}^{T-1} p_{c,j+1}^{w_{c,j}} q_{c,j+1}^{z_{c,j}} \right], \tag{7.3}$$

where $\boldsymbol{p}$ represents the matrix of capture probabilities, $\boldsymbol{\phi}$ the matrix of survival probabilities, $\boldsymbol{\kappa}$ the matrix of recovery probabilities and $\alpha_{c,j} = 1$, when $j = c$, and $\alpha_{c,j} = \prod_{i=c}^{j-1} \phi_{c,i}$ when $c + 1 \leq j \leq K$. The term $\chi_{j,c}$ is the probability that an animal from cohort c alive at t_j is not observed after time t_j, either alive or dead, and it is given by the recursion

$$1 - \chi_{c,j} = (1 - \phi_{c,j})\kappa_{c,j} + \phi_{c,j}(1 - q_{c,j+1}\chi_{c,j+1}),$$

with

$$\chi_{c,K} = 1, \qquad \text{for } c = 1, \ldots, C.$$

The likelihood is determined by four matrices constructed from the raw data, whose elements are defined below, all relating to animals from cohort c, for $1 \leq c \leq C$.

$w_{c,j}$ = the number of animals that are recaptured at time t_{j+1}, $c \leq j \leq T - 1$;
$z_{c,j}$ = the number of animals that are not recaptured at time t_{j+1} but are encountered later in the study, either alive or dead, $c \leq j \leq T - 1$;
$d_{c,j}$ = the number of animals that are observed dead in the interval (t_j, t_{j+1}), $c \leq j \leq K - 1$.
$v_{c,j}$ = number of animals that are last observed at time t_j, $c \leq j \leq T$.

These matrices form a set of *sufficient* statistics, since they are all that we need to retain from the data in order to evaluate the likelihood—it is not necessary to retain all the individual life histories, and this results in a computational saving when the likelihood is formed. It is also attractive to see how simple multinomial likelihood expressions which arise when there are only recovery or only recapture data, generalize to the case of integrated data. The formulation given here is completely general, and allows a wide range of complex models to be specified, making different assumptions about how various model parameters may vary with time as well as age. Formulation of a likelihood in terms of sufficient matrices extends to the case of multi-site models; see King and Brooks (2003), McCrea et al. (2012a) and King and McCrea (2014).

7.2 Formulation as State-Space Models

Capture re-encounter data arise as a consequence of two separate processes, of survival and capture/recovery. This is formalised by Gimenez et al. (2007) and Royle (2008), who show independently that probability models such as those presented above (Eqs. (7.1), (7.1) and (7.3)) may be written as SSMs. This is structurally pleasing, providing a natural link with the use of SSMs in ecology (Buckland et al. 2004; Thomas et al. 2005). It may also simplify complex methods, such as when capture-re-encounter data are analysed simultaneously with census data, described by SSMs (see Chap. 9). As we discuss below, the Bayesian approach using SSM formulations is easily extended to include random effects at the individual level. The general Bayesian state-space modelling of capture-recapture-recovery data is reviewed by King (2012).

In our discussion of SSMs in this section, instead of using $\mathbf{n}_t$ or $\mathbf{N}_t$ for the state vector we use $z(i,t)$ or $Z(i,t)$ where i refers to *individual* animal i. This new notation emphasizes that the use of animal-specific identifiers (marks) allows one to monitor and model the condition of *individuals* from one time period to the next, which is in contrast to cases discussed previously where the condition and abundance of *groups* of animals are being modelled. We note, however, the vector $\mathbf{n}_t$ is still present in this section as it is the vector of the individual animal variables, e.g.

$$\mathbf{n}_t = \begin{bmatrix} z(1,t) \\ z(2,t) \\ \vdots \\ z(n,t) \end{bmatrix} \tag{7.4}$$

For the cases considered in Sects. 7.2.1, 7.2.2 and 7.2.3, the state pdf $\mathbf{g}_t(\mathbf{n}_t|\mathbf{n}_{t-1})$ is a product of n pdfs, either binomial or multinomial, due to assumptions of independence between animals. One can imagine situations, however, where dependence between individual animals occurs, e.g. survival is density dependent, and the pdf $\mathbf{g}_t$ would no longer be a simple product of n pdfs.

7.2.1 Capture-Recapture Data

For illustration we just consider the case of time-dependent parameters. For the CJS model, the state process is given solely in terms of survival, as follows:

$$z(i,t)|z(i,t-1) \sim \text{Bernoulli}(z(i,t-1)\phi_{t-1}) \quad \text{for} \quad t = f_i + 1, \ldots, T, \tag{7.5}$$

and the observation equation is given by

$$y(i,t)|z(i,t) \sim \text{Bernoulli}(p_t z(i,t)) \tag{7.6}$$

where $z(i,t), i = 1, \ldots, n,\ t = 1, \ldots, T$ are Bernoulli indicator random variables for life histories as in Table 7.4, describing whether or not an individual is alive ($z(i,t) = 1$) or dead ($z(i,t) = 0$). We have $z(i, f_i) = 1$, where f_i is the time of first capture of the ith individual. The time-dependent capture probability is denoted by p_t, and the capture history for the ith individual is given by $\{y(i,t)\}_{t=f_i}^{T}$.

To perform Bayesian inference, the state-space formulation of the CJS model allows a straightforward implementation in `WinBUGS` (Gimenez et al. 2007; Royle 2008), provided on the book website (Sect. 1.2).

Table 7.5 Posterior summary statistics for the CJS model fitted to the European dipper data

	Bayesian analysis	Classical analysis
Parameters	Mean (SD)	MLE (SE)
ϕ_1	0.73 (0.13)	0.72 (0.16)
ϕ_2	0.45 (0.07)	0.44 (0.07)
ϕ_3	0.48 (0.06)	0.48 (0.06)
ϕ_4	0.63 (0.06)	0.63 (0.06)
ϕ_5	0.60 (0.06)	0.60 (0.06)
ϕ_6	0.72 (0.14)	NA (NA)
p_2	0.66 (0.13)	0.70 (0.17)
p_3	0.88 (0.08)	0.92 (0.07)
p_4	0.89 (0.06)	0.91 (0.06)
p_5	0.88 (0.05)	0.90 (0.05)
p_6	0.91 (0.05)	0.93 (0.05)
p_7	0.74 (0.14)	NA (NA)

Posterior means and standard deviations (SD) are displayed. For comparison, we also provide maximum-likelihood estimates (MLE) and corresponding standard errors (SE). NAs are for redundant parameters, which cannot be estimated

To run the code, the array of capture histories `y`, the vector `f` of dates of first capture for all individuals, and the constants `n` and `T` for the number of individuals and of capture occasions respectively must be input. As usual in `WinBUGS`, the code is split into two main components, the likelihood and the priors. For the likelihood, the first two lines specify that the initial state of each individual is fixed and alive (at first capture). The conditional distributions of `z` (state equations) and `y` (observation equations) are those specified in Eqs. (7.5) and (7.6) respectively. We assume U(0,1) priors for both time-dependent survival `phi` and detection probabilities `p`.

After discarding a burn-in of 5,000 samples, 50,000 Monte Carlo draws were generated from the posterior distribution. The resulting posterior summary statistics for the European dipper data of Table 7.3 are shown in Table 7.5. For comparison, we also provide the parameter estimates for the CJS model obtained by maximizing the likelihood in Eq. (7.2). Both methods of inference gave essentially the same estimates and measures of precision, with the exception that separate maximum-likelihood estimates of ϕ_6 and p_7 do not exist (see Sects. 5.2.1 and 5.2.6). The Bayesian analysis gives separate estimates of ϕ_6 and p_7 as a consequence of the prior information.

7.2.2 Multi-State Capture-Recapture Models

An important application of capture-recapture models arises when individuals may live in different states or sites (Arnason 1973; Schwarz et al. 1993), and the state-space formulation of the last section extends simply to this case (Dupuis 1995). For illustration, we follow Gimenez et al. (2007) and consider the case of two states. Note that in the standard (non-SSM), description of multi-state capture-recapture models, the state "dead" is never considered, but it does have to be present in the state-space formulation. Let $Z(i,t)$ denote the state vector, which now takes the values $(1,0,0)$, $(0,1,0)$ and $(0,0,1)$ if, at time t, individual i is alive in state 1, 2 or dead respectively. Let $Y(i,t)$ be the corresponding observation vector, taking values $(1,0,0)$, $(0,1,0)$ and $(0,0,1)$ if, at time t, individual i is encountered in state 1, 2 or not encountered, respectively. The model parameters are now $\phi_{i,t}^{(rs)}$, the probability that an animal i survives to time $t+1$ given that it is alive at time t and makes the transition between states r and s over the same interval ($r, s = 1, 2$), as well as $p_{i,t}^{(r)}$ the probability of detecting individual i at time t in state r ($r = 1, 2$). A state-space formulation for the multistate model is then given below, in terms of vector and matrix multiplication:

$$Z(i,t)|Z(i,t-1) \sim \text{multinomial}\left(1, Z(i,t-1)\begin{bmatrix} \phi_{i,t-1}^{(11)} & \phi_{i,t-1}^{(12)} & 1-\phi_{i,t-1}^{(11)} - \phi_{i,t-1}^{(12)} \\ \phi_{i,t-1}^{(21)} & \phi_{i,t-1}^{(22)} & 1-\phi_{i,t-1}^{(21)} - \phi_{i,t-1}^{(22)} \\ 0 & 0 & 1 \end{bmatrix}\right) \tag{7.7}$$

$$Y(i,t)|Z(i,t) \sim \text{multinomial}\left(1, Z(i,t)\begin{bmatrix} p_{i,t}^{(1)} & 0 & 1-p_{i,t}^{(1)} \\ 0 & p_{i,t}^{(2)} & 1-p_{i,t}^{(2)} \\ 0 & 0 & 1 \end{bmatrix}\right) \tag{7.8}$$

where Eqs. (7.7) and (7.8) are the state and observation equations respectively, generalizing Eqs. (7.5) and (7.6).

This formulation can be extended to produce a state-space formulation for integrated recovery and recapture data (Servanty et al. 2010; King 2012) and can also incorporate age effects (Dupuis et al. 2002; King and Brooks 2003; Zheng et al. 2007).

7.2.3 Ring-Recovery Data

In the case of recovery data alone, for the same hidden process that produces the $Z(i,t)$, we can define new Bernoulli random variables $X(i,t)$, taking the value 1 if the ith individual is recovered dead at time t and 0 otherwise. Thus the pattern of 2's and 0's in the life history is readily obtained by adding 1 to the 1's in $X(i,t)$. We can then write the observation equation as

$$Y(i,t)|Z(i,t), Z(i,t-1) \sim \text{Bernoulli}\left((Z(i,t-1) - Z(i,t))\kappa_t\right), \tag{7.9}$$

where κ_t is a time-varying recovery probability.

7.2.4 Extensions

The multi-state formulation has similarities with the multi-event modelling of Pradel (2005), who used hidden-Markov models to extend multi-state models to cope with uncertainty in state assignment, which is a defining characteristic of multi-event models; see also King and McCrea (2014). The SSM formulation of this chapter is well-suited to a Bayesian analysis, and both Gimenez et al. (2007) and Royle (2008) give `WinBUGS` programs for Bayesian inference via Markov chain Monte Carlo (see the dipper code on the book website for an example); see also King et al. (2009) and Schofield et al. (2009). An attraction of the Bayesian approach is that random effects can be easily included in the models, to account for individual variation in parameters such as survival (Royle 2008; see however Gimenez and Choquet 2010 for a classical approach) as well as correlations among parameters at the individual level (Buoro et al. 2010). A comprehensive review is given by King (2012), who also explains how data augmentation may be used for analysis. A characteristic of the CJS model is that it conditions on the first capture of animals. If the model includes age-dependence, and a random effect is also included to account for heterogeneity of survival, then any variation identified in survival might relate to the fact that animals with higher survival probability will be more likely to be sampled than animals with lower survival probability. See Royle (2008) for discussion of this. He analysed the dipper data and identified evidence for heterogeneity of recapture.

If independent census data are also being described by a SSM, as was the case in Besbeas et al. (2002), see Chap. 9, then a combined likelihood can be formed, both components of which are SSMs, which may in some cases be advantageous (Gimenez et al. 2007).

7.3 Discussion and Additional Work

The statistical analysis of mark-re-encounter data has a history stretching back over the past 100 years. The area remains a vibrant one, as new methodology is being developed to cope with developments such as the demands of long-term data sets and the requirements of new marking technologies. For more detail, see McCrea and Morgan (2014) and King (2014). In this short review, we have only described the basic models and procedures, and we list below a number of recent advances.

The use of appropriate covariates to simplify time-dependency in model parameters dates from North and Morgan (1979). Gimenez et al. (2006a) introduce splines for the flexible incorporation of covariates into models. These methods have been adapted to deal with individual covariates (Gimenez et al. 2006b, 2009a). Procedures for dealing with time-varying covariates with missing values are described in Catchpole et al. (2008) and comparisons are provided in Bonner et al. (2010). A hidden-Markov approach is described in Langrock and King (2013). Complex models including age-classes for survival, covariates, including effects of senescence, are described in Catchpole et al. (2000, 2004) and King et al. (2006).

New goodness-of-fit procedures are being developed by McCrea et al. (2014) and McCrea et al. (2012a). Model-selection using score tests is the subject of McCrea and Morgan (2011); see Sect. 5.4.1.2. This is an alternative to methods based on information criteria; see for example Lebreton et al. (1992). Buckland et al. (1997) propose a non-Bayesian method for model-averaging; see Sect. 5.5.1.

Bayesian developments are comprehensively described in King et al. (2009). Grosbois et al. (2009) provide an extension to deal with spatial correlation in capture-recapture, and Lahoz-Monfort et al. (2011) use this methodology for multi-species systems. Both papers incorporate random effects and use a Bayesian formulation and analysis.

Cole et al. (2010) present new ways of determining the parametric structure of complex probability models, which is particularly relevant to mark-re-encounter models, when it may not be possible to estimate all of the model parameters. A simple case of this arises in the CJS model, as can be seen in Table 7.5; see Chap. 5. These methods are applied to ring-recovery models in Cole et al. (2012), multi-state models in Cole (2012), and capture-recapture-recovery models in Hubbard et al. (2014).

Chapter 8
Estimating Abundance from Mark-Recapture Data

In terms of modelling population dynamics, the mark-recapture literature has in recent years been dominated by methods for estimating survival, as described in Chap. 7. In this chapter, we consider open-population mark-recapture methods for estimating abundance, survival and births. We first summarise conventional methods (Seber 1973, 1982). In these, abundance is estimated largely as for closed-population mark-recapture methods (Borchers et al. 2002:104–130), except that the number of marked animals surviving to the next capture occasion must be estimated. Survival is estimated as in Chap. 7. Finally, number of births is usually estimated by subtracting the estimated number of surviving animals from the previous time period from the estimated population size in the current time period.

The shortcoming of the conventional approach is that there is no embedded population dynamics model. Thus estimated birth rates can be biologically impossible; for example, estimated number of births may exceed the number of breeding females for species that give birth to at most one young per year. In fact, the above process of estimating numbers of births by subtraction often leads to negative estimates of numbers of births. In Sect. 8.2, we provide state-space formulations of mark-recapture models that ensure that estimated numbers of births and estimated changes in the population are consistent with biological reality.

In common with other chapters, we assume that the basic unit of time is one year, so that there is one capture session per year. The methods are readily generalizable to different time units.

K.B. Newman et al., *Modelling Population Dynamics: Model Formulation, Fitting and Assessment using State-Space Methods*, Methods in Statistical Ecology,
DOI 10.1007/978-1-4939-0977-3_8,

8.1 Conventional Mark-Recapture Models for Open Populations

8.1.1 A Likelihood Formulation

In the following, we assume that year t ends with a sampling occasion, and we assume that all animals caught are released alive with permanent marks.

Let

N_t = number of animals in the population just before the sampling occasion in year t, $t = 1, \ldots, T$;
M_t = number of marked animals in the population just before the sampling occasion in year t;
$V_t = N_t - M_t$ = number of unmarked animals in the population just before the sampling occasion in year t;
s_t = number of animals caught during the sampling occasion in year t;
m_t = number of marked animals caught during the sampling occasion in year t;
$v_t = s_t - m_t$ = number of unmarked animals caught during the sampling occasion in year t, and marked before release;
r_t = number of marked animals released in year t and subsequently recaptured;
z_t = number of animals caught before year t that are not recaptured in year t, but are subsequently;
χ_t = probability that an animal released in year t is not recaptured.

Note that we use s_t rather than the more standard n_t for sample size in year t, to avoid confusion with the state vector $\mathbf{n}_t$. Similarly, we use V_t and v_t to indicate unmarked animals in the population and sample respectively in year t rather than the more standard U_t and u_t to avoid confusion with intermediate state vectors $\mathbf{u}_t$.

If we assume that every animal in the population has the same probability of capture p_t in year t, and the same probability of survival ϕ_t from the end of year t to the end of year $t+1$, then we might assume a multinomial model, giving the following likelihood apart from constants (Seber 1965):

$$\prod_{t=1}^{T} \frac{V_t!}{(V_t - v_t)!} p_t^{s_t} (1 - p_t)^{z_t + V_t - v_t} \prod_{t=1}^{T-1} \phi_t^{r_t + z_t} \chi_t^{s_t - r_t}. \tag{8.1}$$

If for the above model, ϕ_t and p_t are allowed to vary by year only, we obtain the Jolly-Seber model (Jolly 1965; Seber 1965). The likelihood can be extended to allow deaths on capture (Jolly 1965) or known deaths between capture occasions (Buckland 1980). If either survival probabilities or capture probabilities or both are assumed to be constant over years, we have the reduced-parameter models of Jolly (1982).

By modelling ϕ_t and/or p_t as functions of covariates, more biologically relevant models may be obtained (Pollock 2002). Thus for example, adopting a similar model to that proposed in Sect. 2.2.1, we might model ϕ_t as

$$\phi_t = \frac{1}{1 + \exp\{\beta_0 + \beta_1 r_{t+1} + \beta_2 N_t\}} \tag{8.2}$$

where r_{t+1} is an environmental covariate and N_t is abundance at the end of year t.

If covariates are measured on individual animals, the likelihood may be readily extended to provide individual-based models, similar to the closed-population methods of Huggins (1989) and Alho (1990), and to the methods of Morgan et al. (2004) (see Chap. 7) for estimating survival rates from mark-recapture data. Of course, such covariates can usually only be recorded for captured animals.

The model parameters may also be modelled as random effects, as was done for closed populations by Coull and Agresti (1999), offering even greater flexibility. For example, we might model ϕ_t as:

$$\phi_t = \frac{1}{1 + \exp\{\beta_0 + \beta_1 r_{t+1} + \epsilon_{t+1}\}} \tag{8.3}$$

where $\epsilon_{t+1} \sim \text{normal}(0, \sigma^2)$.

These strategies allow heterogeneity amongst animals in the probability of capture or survival to be modelled. Time-varying covariates are problematic because typically, these can only be recorded when animals are caught. Methods for handling the missing values have been developed by Catchpole et al. (2008), Bonner et al. (2010) and Langrock and King (2013).

Pollock (1982) proposed a different strategy for modelling heterogeneity in capture probabilities. At that time, methods were available for modelling such heterogeneity for closed populations, but not for open populations. He therefore suggested a design in which a closed-population mark-recapture study is conducted over a short period each year, to allow robust estimation of population size. The Jolly-Seber survival estimators, treating each year as a single sample (i.e. pooling data within a year), are still adopted. A slight modification of the Jolly-Seber birth estimators is now used, in which the annual abundance estimates are obtained from the within-year closed-population studies using a model that allows heterogeneity in the capture probabilities.

8.1.2 Modelling Births

Note that Eq. (8.1) is deterministic with respect to births, in that the number of births B_t in year t is taken to be the population size at the end of year $t + 1$, N_{t+1}, less the number of survivors $N_t\phi_t$ from the end of year t, so that:

$$\hat{B}_t = \hat{N}_{t+1} - \hat{N}_t\hat{\phi}_t \ . \tag{8.4}$$

Crosbie and Manly (1985) explicitly included the birth process in the likelihood, overcoming one of the limitations in the Jolly-Seber formulation. They achieved this by defining N to be the total number of unique animals that existed in the population in the period $[1, T]$. A multinomial distribution is assumed to model how these animals split into the possible capture histories, and for each capture history, a component of the likelihood is defined for each time interval that an animal could have entered the population (i.e. birth), and for each time interval that an animal could have died. For example, if there were $T = 6$ sampling occasions, and a particular animal was recorded in years 2, 3 and 5, then the non-zero contributions to the likelihood correspond to birth before $t = 1$ or birth in the interval $(1, 2)$, followed by death in the interval $(5, 6)$, or death after $t = 6$. Schwarz and Arnason (1996) developed this approach further.

Link and Barker (2005) adopted a similar strategy, but introduced birth rate parameters into their formulation, to replace the less biologically relevant multinomial probabilities of the above approach. They also used a hierarchical modelling approach to model birth and survival rates, and used Markov chain Monte Carlo methods to fit their models.

However, the above approaches still fail to respect biological reality. For example, if adults can have no more than a single young in a year, the above methods do not ensure that this constraint is respected; there is no embedded population dynamics model. We now consider state-space models that remove this limitation.

8.2 State-Space Formulations

Equations (3.3)–(3.5) gave the state and observation pdf's for a state-space model:

$$\text{Initial state pdf}: g_0(\mathbf{n}_0|\boldsymbol{\theta}) \tag{8.5}$$

$$\text{State } t \text{ pdf}: g_t(\mathbf{n}_t|\mathbf{n}_{t-1}, \boldsymbol{\theta}) \tag{8.6}$$

$$\text{Observation } t \text{ pdf}: f_t(\mathbf{y}_t|\mathbf{n}_t, \boldsymbol{\psi}). \tag{8.7}$$

The demographic processes such as birth and death determine the numbers of animals in each state. The state pdf $g_t(\mathbf{n}_t|\mathbf{n}_{t-1}, \boldsymbol{\theta})$ represents the variability arising from these demographic processes in updating from year $t - 1$ to year t. In mark-recapture, there is also the stochastic process that determines which animals are caught each year. We can represent this process in the observation pdf. However, at a minimum, we must define different states corresponding to whether animals are marked or not. Thus unusually, the processes that generate the states are not independent of the observation process (but see conditionally Gaussian modelling in Sect. 9.7.1). If we adopt this approach, then for $t \geq 2$, we should replace Eq. (8.6) by

$$g_t(\mathbf{n}_t|\mathbf{n}_{t-1}, \mathbf{y}_{t-1}, \boldsymbol{\theta}). \tag{8.8}$$

A second option is to define the capture process to be another component of the state process. In this case, we assign animals with different capture histories to different states. That is, an animal's capture history determines its state (along with other relevant variables such as sex and age or size class). In this case, as the stochastic capture process is also incorporated into the state pdf, the observation pdf becomes degenerate (unless there are additional data on the population over and above the mark-recapture data). We consider both options below. We term the first option "unconditional" and the second option "conditional".

In the conditional approach, parameters relating to probability of capture appear in the vector of state parameters, $\boldsymbol{\theta}$. In the unconditional approach, the capture histories are the observations, and do not influence the state of an animal. Consequently, parameters related to capture probability appear as observation model parameters, $\boldsymbol{\psi}$. We term the first approach a conditional approach because for a given year, we condition on the known numbers of animals with capture histories that include capture in that year. Thus there is no observation error, and stochasticity enters through the uncertainty in the numbers of animals by state with capture histories that do not involve capture in the current year. With the unconditional approach, we treat capture as a stochastic observation process, and model the capture histories as if they are just one random realisation of that process. The unconditional approach is more consistent with the traditional mark-recapture methods summarized above. It should be noted that the two approaches lead to different fitting algorithms, but the model being fitted is the same either way.

Jolly (1965) termed quantities such as N_t "conditional parameters". In the following formulations, whether conditional or unconditional, we adopt a Bayesian approach, for which the posterior distribution for N_t obtained reflects uncertainty arising from the observation process only. If we wish to draw inference on $E[N_t]$ where expectation is across all possible realisations from an underlying superpopulation, we need to simulate a sample of these realisations, having obtained samples from the posterior distributions of the demographic parameters such as survival and birth rates.

8.2.1 The Unconditional Approach

To fit a state-space model, we must specify models for the initial state, state t and observation t pdfs of Eqs. (8.5), (8.8) and (8.7). For a Bayesian approach, we must also specify the prior distribution $\pi(\boldsymbol{\theta}, \boldsymbol{\psi})$. For the unconditional approach, the state pdfs are specified in the standard way; the dependence of the state pdf in year t on the observations in year $t-1$ is not problematic. Because the state model handles numbers of animals by state in the population, the observation pdf is simpler than the likelihood for conventional mark-recapture methods. For example, if animals are grouped into just two states corresponding to marked and unmarked, the likelihood given by Eq. (8.1) can be replaced by the observation pdf:

$$f(m_1, \cdots, m_T, v_1, \cdots, v_T | M_1, \cdots, M_T, V_1, \cdots, V_T, \psi) = \prod_{t=1}^{T} \binom{M_t}{m_t} p_t^{m_t} (1-p_t)^{M_t - m_t} \times \binom{V_t}{v_t} p_t^{v_t} (1-p_t)^{V_t - v_t} \tag{8.9}$$

where $p_t \equiv p_t(\psi)$ and $M_1 = m_1 = 0$. We do not need a pdf for survival, or indeed for births, since these appear in the state model.

We note that this observation pdf is a product of components, each representing one year; these components are the observation t pdfs of Eq. (8.7). Thus $f_t(\mathbf{y}_t|\mathbf{n}_t, \psi) = f_t(m_t, v_t|M_t, V_t, \psi) = \binom{M_t}{m_t} p_t^{m_t}(1-p_t)^{M_t-m_t} \times \binom{V_t}{v_t} p_t^{v_t}(1-p_t)^{V_t-v_t}$. Fitting now proceeds using any of the algorithms of Chap. 4.

This separation of the capture process from demographic processes both simplifies the observation pdf, which now reflects the capture process only, and allows us to specify biologically realistic models for the demographic processes via the state process pdf. Schofield and Barker (2008) adopted this strategy to assess evidence for density dependence in birth and survival rates, and Schofield et al. (2009) showed how WinBUGS may be used to fit such models. Bishop (2009) compared this approach with the conditional approach (below).

In the above formulation, the observation vector $\mathbf{y}_t$ has just two elements, corresponding to the number of marked animals m_t and unmarked animals v_t caught in year t. Similarly, the state vector $\mathbf{n}_t$ has two elements, corresponding to the number of marked animals M_t and unmarked animals V_t in the population just before the sampling occasion in year t. Now suppose we have K states, with $\mathbf{n}_t = (n_{1,t}, \cdots, n_{K,t})'$ and $\mathbf{y}_t = (m_{1,t}, \cdots, m_{K,t}, v_{1,t}, \cdots, v_{K,t})'$. Then the observation pdf is given by

$$\prod_{t=1}^{T} f_t(\mathbf{y}_t|\mathbf{n}_t, \psi) \tag{8.10}$$

where

$$f_t(\mathbf{y}_t|\mathbf{n}_t, \psi) = \prod_{k=1}^{K} \binom{M_{k,t}}{m_{k,t}} p_{k,t}^{m_{k,t}} (1 - p_{k,t})^{M_{k,t} - m_{k,t}} \times \binom{V_{k,t}}{v_{k,t}} p_{k,t}^{v_{k,t}} (1 - p_{k,t})^{V_{k,t} - v_{k,t}} \tag{8.11}$$

and $p_{k,t}$ is the probability of capture in year t of an animal in state k. Various models can be considered for this probability. For example, it could be assumed independent of state k or of year t, or both. It could also be modelled as a function of covariates (which could include previous capture history), or using a random effects model (similar to Eq. (8.3) for modelling survival).

8.2.2 *The Conditional Approach*

In the conditional approach, there are no observation parameters $\boldsymbol{\psi}$, and the observation pdf is degenerate; for year t, we observe the number of animals in states corresponding to capture in that year without error. The uncertainty in year t now relates to states corresponding to animals that are not caught in that year, and this is modelled through the state t pdf, $g_t(\mathbf{n}_t|\mathbf{n}_{t-1}, \boldsymbol{\theta})$. The state vector $\mathbf{n}_t$ expands each year, as the number of possible capture histories increases. The requirement to condition on numbers of animals with capture histories corresponding to capture in year t complicates model fitting.

Suppose initially an animal's state is determined solely by its capture history. Let

$\mathbf{s}_t$ = vector of numbers of animals for each capture history in $1, \ldots, t$ that includes capture in year t;

$\mathbf{v}_t$ = vector of numbers of animals for each capture history in $1, \ldots, t$ that excludes capture in year t.

$\mathbf{n}_t = \begin{bmatrix} \mathbf{s}_t \\ \mathbf{v}_t \end{bmatrix}$ = vector of numbers of animals with each possible capture history up to and including captures in year t;

Thus $\mathbf{s}_t$ is observed, whereas $\mathbf{v}_t$ is not. The state t pdf is still written $g_t(\mathbf{n}_t|\mathbf{n}_{t-1}, \boldsymbol{\theta})$, where $\boldsymbol{\theta}$ now contains capture probability parameters in addition to population parameters. The (degenerate) observation t pdf is now $f_t(\mathbf{y}_t|\mathbf{n}_t) \equiv f_t(\mathbf{s}_t|\mathbf{s}_t) = 1$ if $\mathbf{s}_t = \mathbf{y}_t$ and is zero otherwise. Stochastic variation in $\mathbf{n}_t$ is now restricted to those elements that are also in $\mathbf{v}_t$. Fitting algorithms thus need to handle the conditioning on the entire time series represented by $\mathbf{s}_t$, $t = 1, \ldots, T$. Bishop (2009) considers this method, outlines an approach for simulating populations that respect the constraints imposed by the observed capture histories, and discusses the problems associated with this approach.

If for each capture history, an animal can belong to one of K states, then in year t, each state can separately be modelled in the above way.

To fit a model in this framework, we need to specify a model for capture. If this model is the same as for the unconditional method, then this is just a different way to fit the same model. For example the model underlying the likelihood of Eq. (8.9) is one that assigns the same probability of capture to every animal and assumes independence across years. We can use the same model to define the state pdf for the conditional approach.

8.3 A Simple Example

We use an example of a mark-recapture survey of a population of animals with two age classes, from Buckland et al. (2004). We model the female population only. Animals have a constant survival rate ϕ_j in their first year and an annual survival rate ϕ_a subsequently. Mortality occurs primarily in the winter and spring. Breeding first occurs at age 1 and the number of births per adult is distributed as a Bernoulli random variable with probability of success ρ.

Births occur in summer and are followed by a single sample of a mark-recapture experiment in the autumn. It is assumed that there is no mortality in the period between births and the mark-recapture sample, although the model could be extended (by adding another sub-process) to include such mortality. Juvenile animals can be distinguished from adults on capture. In year t, juveniles are caught with probability p_{jt} and adults with probability p_{at}, $t = 1, \ldots, T$. (Age-dependent probabilities of capture commonly arise in practice, for example because young animals are inexperienced and more prone to capture.)

The state vector $\mathbf{n}_t$ contains the abundances of animals, with separate enumeration of marked and unmarked animals, immediately before the mark-recapture experiment in year t. Although conventional mark-recapture formulations tally capture histories of marked animals, both alive and dead, we only record a capture history for an animal while it is alive. This is because the state vector at time t only includes animals alive at time t. A capture history of "110" therefore indicates an animal that was caught in its first year of life and again in its second year of life, survived into its third year, but was not caught in that year.

Given the difficulties in implementing the conditional approach (Bishop 2009), we consider here only the unconditional approach. The length of $\mathbf{n}_t$ for $t > 1$ is three, corresponding to marked ($M_{a,t}$) and unmarked ($V_{a,t}$) adults and unmarked juveniles ($V_{j,t}$), while $\mathbf{n}_1$ is of length two, as there are no marked animals before the first capture session ($M_{a,1} = 0$); note that $M_{j,t} = 0$ also, as no juveniles can have been marked ahead of the capture session in year t. The state vector for time period t is thus

$$\mathbf{n}_1 = \begin{bmatrix} V_{j,1} \\ V_{a,1} \end{bmatrix} \qquad \mathbf{n}_t = \begin{bmatrix} V_{j,t} \\ M_{a,t} \\ V_{a,t} \end{bmatrix}.$$

The observation vector $\mathbf{y}_t$ may be reduced to two elements for $t = 1$ and three elements for $t \geq 2$ (no juveniles can be already marked, so $m_{j,t} = 0$):

$$\mathbf{y}_1 = \begin{bmatrix} v_{j,1} \\ v_{a,1} \end{bmatrix} \qquad \mathbf{y}_t = \begin{bmatrix} v_{j,t} \\ m_{a,t} \\ v_{a,t} \end{bmatrix}.$$

Three sub-processes, two stochastic and one deterministic (advancement of juveniles to either the young or the adult stage and advancement of young to adult), generate $\mathbf{n}_t$ from $\mathbf{n}_{t-1}$. The sequencing is winter/spring survival (following mark and recapture), advancement, and births. The initial state vector consists of juveniles and adults in the autumn, before the first marking.

For time period t, the processes are modelled as follows. We must first allow for the newly-marked animals, which reflects the dependence of the state pdf on the observations of the previous year. Thus the marked population is supplemented by the number of animals newly-marked, while the same number is subtracted from the unmarked population. Then comes winter/spring survival $\mathbf{u}_{1,t} \sim H_{1,t}(\mathbf{n}_{t-1})$:

$$\begin{bmatrix} u_{1,j,m,t} & \sim \text{Binomial}(v_{j,t-1}, \phi_j) \\ u_{1,j,v,t} & \sim \text{Binomial}(V_{j,t-1} - v_{j,t-1}, \phi_j) \\ u_{1,a,m,t} & \sim \text{Binomial}(M_{a,t-1} + v_{a,t-1}, \phi_a) \\ u_{1,a,v,t} & \sim \text{Binomial}(V_{a,t-1} - v_{a,t-1}, \phi_a) \end{bmatrix}.$$

Then age incrementation of juveniles, $\mathbf{u}_{2,t} \sim H_{2,t}(\mathbf{u}_{1,t})$:

$$\begin{bmatrix} u_{2,a,m,t} & = u_{1,j,m,t} + u_{1,a,m,t} \\ u_{2,a,v,t} & = u_{1,j,v,t} + u_{1,a,v,t} \end{bmatrix}.$$

Then births, $\mathbf{n}_t \equiv \mathbf{u}_{3,t} \sim H_{3,t}(\mathbf{u}_{2,t})$:

$$\begin{bmatrix} u_{3,j,t} & \sim \text{Binomial}(u_{2,a,m,t} + u_{2,a,v,t}, \rho) \\ u_{3,a,m,t} & = u_{2,a,m,t} \\ u_{3,a,v,t} & = u_{2,a,v,t} \end{bmatrix}.$$

The conditional expectation for $\mathbf{n}_t$ can be expressed as products of matrices:

$$\text{E}_{\mathbf{n}_t|\mathbf{n}^*_{t-1}}[\mathbf{n}_t] = B_t A_t S_t \mathbf{n}^*_{t-1}$$

where

$$\mathbf{n}^*_{t-1} = \begin{bmatrix} v_{j,t-1} \\ V_{j,t-1} - v_{j,t-1} \\ M_{a,t-1} + v_{a,t-1} \\ V_{a,t-1} - v_{a,t-1} \end{bmatrix} \qquad S_t = \begin{bmatrix} \phi_j & 0 & 0 & 0 \\ 0 & \phi_j & 0 & 0 \\ 0 & 0 & \phi_a & 0 \\ 0 & 0 & 0 & \phi_a \end{bmatrix}$$

$$A_t = \begin{bmatrix} 1 & 0 & 1 & 0 \\ 0 & 1 & 0 & 1 \end{bmatrix} \qquad B_2 = \begin{bmatrix} \rho & \rho \\ 1 & 0 \\ 0 & 1 \end{bmatrix}.$$

Note that $\mathbf{n}^*_{t-1}$ incorporates dependence of the expectation on both $\mathbf{n}_{t-1}$ and $\mathbf{y}_{t-1}$. The component of the likelihood corresponding to year 1 is

$$f_1(\mathbf{y}_1|\mathbf{n}_1, \boldsymbol{\psi}) = \binom{V_{j,1}}{v_{j,1}} p_{j,1}^{v_{j,1}}(1 - p_{j,1})^{V_{j,1}-v_{j,1}} \times \binom{V_{a,1}}{v_{a,1}} p_{a,1}^{v_{a,1}}(1 - p_{a,1})^{V_{a,1}-v_{a,1}} \tag{8.12}$$

and for year $t \geq 2$, given that no juveniles can have been marked previously (i.e. $M_{j,t} = m_{j,t} = 0$) we have

$$
\begin{aligned}
f_t(\mathbf{y}_t|\mathbf{n}_t, \boldsymbol{\psi}) &= \binom{V_{j,t}}{v_{j,t}} p_{j,t}^{v_{j,t}} (1 - p_{j,t})^{V_{j,t} - v_{j,t}} \\
&\quad \times \binom{M_{a,t}}{m_{a,t}} p_{a,t}^{m_{a,t}} (1 - p_{a,t})^{M_{a,t} - m_{a,t}} \times \binom{V_{a,t}}{v_{a,t}} p_{a,t}^{v_{a,t}} (1 - p_{a,t})^{V_{a,t} - v_{a,t}} \\
&= \binom{V_{j,t}}{v_{j,t}} \binom{M_{a,t}}{m_{a,t}} \binom{V_{a,t}}{v_{a,t}} p_{j,t}^{v_{j,t}} (1 - p_{j,t})^{V_{j,t} - v_{j,t}} p_{a,t}^{m_{a,t} + v_{a,t}} (1 - p_{a,t})^{M_{a,t} + V_{a,t} - (m_{a,t} + v_{a,t})}.
\end{aligned}
\tag{8.13}
$$

8.4 Discussion

The advantages of a state-space approach to modelling open-population mark-recapture data are clear. The underlying processes of survival and birth are an integral part of the model, and are not simply estimated empirically. This ensures that estimated birth rates are biologically plausible, and that estimated abundances do not fluctuate unrealistically. There is potential for extending the models, for example to include movement rates between populations in a metapopulation, or to incorporate species interactions in a predator-prey system or in a community of competing species. Further, specification of the observation likelihood is simplified, because the population processes are removed to the state pdf's. It is also straightforward to incorporate other components in the observation likelihood, if other types of data are recorded in addition to the mark-recapture data. However, this is a field in need of further research.

The methods presented here are also useful for assessing the effects of a management intervention. For example, if a scheme is introduced, compensating farmers for implementing conservation measures expected to favour the species of interest, an indicator variable can be introduced, taking the value zero for animals captured outside of areas operating the scheme, and one for those captured inside. This indicator variable can then be included as a covariate in the survival and/or birth models.

Open-population mark-recapture studies for estimating abundance of vertebrate populations are much less common than ones for estimating survival. In part, this is due to the difficulty of modelling heterogeneity in the capture probabilities. There are now better tools available for modelling the heterogeneity, and the difficulty can be reduced further or eliminated if there are supplementary studies for estimating abundance. Pollock (1982) used closed-population mark-recapture studies within each time period, and modelled heterogeneity in the corresponding capture probabilities. Another possibility is to conduct independent distance sampling surveys in each time period.

Chapter 9
Integrated Population Modelling

9.1 Introduction

In Chap. 5, we recommended that formulation of population dynamics models should be guided by aims to answer specific scientific questions or assess or predict the effects of management actions. Management actions might target a specific life stage. For example, we might ask "How does removing wetland plants (such as bulrush or cattail) that have started to cover ponds and reduce the amount of open water in a waterfowl breeding area affect reproductive success?" The consequences of actions, however, typically ripple throughout the entire population life history and effective management requires more detailed ecological study. This in turn requires information about demographic processes and abundances for *multiple* life stages to characterize the population dynamics.

Information at the population and individual levels is often simultaneously available in monitoring programmes of wildlife populations. For example one survey might be designed to provide information about survival for a specific life stage, another for reproductive success, and a third for total population abundance. Often the information from each survey is analysed in isolation, and the separate results are used to fill the elements of a Leslie or Lefkovitch population projection matrix (Chap. 2). However, the surveys may provide overlapping information about demographic processes or abundances for multiple life stages and analyses that utilize that overlap are likely to be more powerful and provide more information than multiple piecemeal analyses.

In this chapter, we discuss approaches that do utilize the overlapping information, namely integrated population modelling (IPM). We define IPM to be the fitting of a population dynamics model to two or more sources of data where (i) the fitting is done in a *single or simultaneous* stage, and (ii) each source provides information at either the population or individual level. A third common feature of IPM, but not necessary to our definition, is (iii) at least two sources provide overlapping information about one or more population processes. A review of applications of integrated population modelling is provided by Schaub and Abadi (2011).

K.B. Newman et al., *Modelling Population Dynamics: Model Formulation, Fitting and Assessment using State-Space Methods*, Methods in Statistical Ecology,
DOI 10.1007/978-1-4939-0977-3_9,

We emphasise *single or simultaneous* to distinguish IPM from multi-stage or sequential fitting procedures. An example of a multi-stage fitting procedure is to use one data set to estimate survival probabilities for a sequence of years, another to estimate reproductive success, and then a third to estimate annual abundance estimates, where the three sets of estimates are calculated independently of one another, and the three sets of estimates are then used to fit a population dynamics model. In contrast, an IPM analysis uses all three data sets in a single combined analysis to fit the population dynamics model with simultaneous estimation of survival, reproduction and abundance. Maunder (1998) discusses several advantages of IPMs over multi-stage (non-simultaneous) analyses and we highlight some of these advantages and others below.

An IPM analysis can account for correlations between survival, reproduction and abundances. For example, years of higher survival are often years of higher reproductive success. Parameters shared by multiple data sets can be estimated with greater precision, or at least the uncertainty in an estimate based on combined data sets can be more coherently estimated, and the sharing of parameters can overcome parameter redundancy problems (Sect. 5.2). The gray whale analysis in Sect. 6.4.2 is an example of an IPM where non-identifiability problems were mitigated to a degree; abundance estimates alone make separation of survival and reproduction rates difficult at best, but the addition of harvest data provided additional information about both sets of parameters and improved estimability.

The availability of multiple surveys can also influence model formulation (Sect. 5.1), allowing for finer temporal and spatial modelling of population dynamics when for example different surveys are taken at different times of the year.

In keeping with the theme of the book, we first show how state-space models can provide a unifying framework for readily integrating data from multiple sources. However, such a completely unified SSM approach to analysing multiple data sources can be technically intricate and daunting. Most of this chapter focuses on an alternative IPM approach, an approach we call a connected likelihood approach. This latter approach can be less technically demanding, where a SSM is combined with other non SSM models for different data sets. The connected likelihood approach builds naturally on methods for estimating survival probabilities (Chap. 7) and abundances (Chap. 8) over time. In Chap. 8, we focussed on methods for estimating animal abundance from mark-recapture data. However, population abundance measures or estimates can be provided by other kinds of data and surveys, such as line transect surveys or aerial counts for randomly selected plots. Such surveys in general are aimed at detecting trends, such as growth or decline, in total population numbers, often at large scale, e.g. abundances at a national level. We shall use the term *census* in a generic sense in this chapter for any procedure used for estimating the size of a population or of a predefined part of it from field data. General procedures for estimating the size of a population are discussed in Chap. 6.

The structure of this chapter is as follows. Sect. 9.2 discusses the single SSM approach to IPM mentioned previously. Section 9.3 describes the connected likelihood approach and includes a worked example. An approximation that greatly assists the connected likelihood model is given and illustrated in Sect. 9.4 and technical issues that arise in forming the Kalman filter likelihood are considered in Sect. 9.5. Extending the connected likelihood approach to multi-site and multi-state models is the topic of Sect. 9.6. The chapter ends (Sect. 9.7) with a variety of extensions and issues particular to IPMs, including conditional Gaussian (i.e. normal) modelling (Sect. 9.7.1), Bayesian methods (Sect. 9.7.3), and goodness-of-fit (Sect. 9.7.2).

9.2 Integrated Modelling within an SSM Framework

As discussed in Chap. 5, the temporal, spatial and biological resolution of the available data constrains formulation of the state process model in terms of what state vector components and parameters can be estimated. When more than one survey or data set is available, flexibility in the formulation of the SSM may be greater than what is feasible with a single data set. Multiple surveys can provide information on population abundances distinguished by age, sex, maturity or life history stage. Multiple surveys can also provide information about different population dynamic processes such as survival and reproduction. Data from multiple surveys can be incorporated in the SSM framework in two different ways: either expand the number of components of the observation vector or increase the number of observation vectors.

When multiple surveys make measurements on the population at the same point in time, the observation vector is enlarged to include data from each survey. If the different surveys are measuring the same components of the state vector, the data are simply replicate measurements that may or may not be independent of each other (conditional on the state component) and in general have different variances. For example, a mark-recapture survey and a line transect survey might be carried out within a week of each other and the population is assumed to be relatively static during that week. Estimates of abundance for week t from the two surveys are calculated ($y_{mr,t}$ and $y_{lt,t}$ for mark-recapture and line transect respectively) with corresponding standard errors ($s_{mr,t}$ and $s_{lt,t}$). Assuming that both are unbiased estimates of total abundance N_t, and are independent of each other, a general expression for the observation model is the following.

$$\begin{bmatrix} y_{mr,t} \\ y_{lt,t} \end{bmatrix} \sim D\left(\boldsymbol{\mu}_t = \begin{bmatrix} N_t \\ N_t \end{bmatrix}, \Sigma = \begin{bmatrix} s_{mr,t}^2 & 0 \\ 0 & s_{lt,t}^2 \end{bmatrix}\right)$$

where D is an arbitrary bivariate distribution with expected value vector $\boldsymbol{\mu}_t$, and variance-covariance matrix Σ. In this example, the observations are taken to be derived quantities rather than the raw data (e.g. numbers marked and recaptured in the mark-recapture survey, numbers counted and distances to animals in the line transect survey). The raw data could be the observations, but this often requires considerably more complex distributional structures; e.g. the likelihood models underlying distance sampling (Buckland et al. 2001). Knape et al. (2013) examined the impact of using derived quantities instead of raw data and found that the loss of information was slight in the cases considered (also see Sect. 5.1). Another scenario is that different management agencies carry out separate surveys for distinct, non-overlapping land units. For example, duck surveys are made at three different wildlife refuges at nearly the same point in time. The population total might be spatially partitioned by each refuge, where $N_{A,t}$, $N_{B,t}$ and $N_{C,t}$ are the abundances for refuges A, B and C, respectively. The state vector is expanded accordingly and the components of the observation vector are independent abundance estimates, denoted $y_{A,t}$, $y_{B,t}$, and $y_{C,t}$, which are matched with each refuge's population abundance. For example:

$$\begin{bmatrix} y_{A,t} \\ y_{B,t} \\ y_{C,t} \end{bmatrix} \sim D\left(\boldsymbol{\mu}_t = \begin{bmatrix} n_{A,t} \\ n_{B,t} \\ n_{C,t} \end{bmatrix}, \Sigma = \begin{bmatrix} s^2_{A,t} & 0 & 0 \\ 0 & s^2_{B,t} & 0 \\ 0 & 0 & s^2_{C,t} \end{bmatrix}\right)$$

In the case of multiple surveys making measurements at different points in time and perhaps focussing on different population sub-processes, additional observation vectors could be inserted and paired with different state vectors. For a concrete example, we revisit the BRS model of Sects. 2.3 and 3.2.2. The BRS model had two states (immature and mature animals) and three sub-processes, survival (S), growth (i.e. maturation, R) and birth (B), occurring in that sequence. We assume that three separate surveys were carried out independently to estimate the parameters corresponding to these processes, namely ϕ_1, ϕ_2, π and ρ (see Eqs. (2.2), (2.5) and (2.6)). We also assume that a fourth survey was carried out just after the breeding season, giving an estimate of total abundance $(n_{1,t}, n_{2,t})$; thus these are *census* data as defined previously. For simplicity, assume that the animals sampled to estimate the survival probabilities are then followed throughout the subsequent growth and reproduction processes with perfect detectability, i.e. their growth and reproduction numbers are known without error. The observation sub-processes are modelled as follows.

$$\begin{aligned}
(y_{1,t-1}, y_{2,t-1}) &\sim D\left((n_{1,t-1}, n_{2,t-1}), \theta\right) \\
y_{1(s),1,t} &\sim \text{binomial}\left(n_{1,t-1}, p\phi_1\right) \\
y_{1(s),2,t} &\sim \text{binomial}\left(n_{2,t-1}, p\phi_2\right) \\
y_{2(r),2,t} &\sim \text{binomial}\left(y_{1(s),1,t}, \pi\right) + y_{1(s),2,t} \\
y_{3(b),1,t} &\sim \text{binomial}\left(y_{2(r),2,t}, \rho\right)
\end{aligned}$$

The terms $y_{1,t-1}$ and $y_{2,t-1}$ are the census data with D an arbitrary distribution reflecting the uncertainty in the census data with corresponding parameter θ, while p is the probability of capture for the survival study. If different animals were used for sampling the three sub-processes, then the population abundances at each point in time would be substituted for $y_{1(s),1,t}$ and $y_{2(r),2,t}$ and two additional capture probabilities are added. For example, using the notation from Eqs. (3.25) and (3.26):

$$\begin{aligned}
(y_{1,t-1}, y_{2,t-1}) &\sim D\left((n_{1,t-1}, n_{2,t-1}), \theta\right) \\
y_{1(s),1,t} &\sim \text{binomial}\left(n_{1,t-1}, p_s\phi_1\right) \\
y_{1(s),2,t} &\sim \text{binomial}\left(n_{2,t-1}, p_s\phi_2\right) \\
y_{2'(r),2,t} &\sim \text{binomial}\left(u_{1(s),1,t}, p_r\pi\right) \\
y_{3(b),1,t} &\sim \text{binomial}\left(u_{2(r),2,t}, p_b\rho\right)
\end{aligned}$$

where p_s, p_r, and p_b denote the probability of sampling animals for the three sub-processes and $y_{2'(r),2,t}$ denotes the immature animals that just matured.

Section 9.3 presents an alternative approach to using data from multiple surveys in which the state-space model formulation is just used for a subset of one or more surveys and alternative formulations characterize data from other surveys. In particular Sect. 9.3 constructs different likelihoods for different surveys but two or more likelihoods have one or more parameters in common.

9.3 Integrated Modelling with Connected Likelihoods

Census data are naturally dependent upon animal survival and fecundity, and this observation motivates this section and much of the remainder of this chapter. When information from several sources is available for a particular species, it is natural to consider the extent to which the available types of survey data are compatible and how they can complement each other. In early work, the emphasis was on the former, and matching the different types of analyses was done in an *ad hoc* way, by comparing model-based and census-based population growth rates (Coulson et al. 2001) or by checking visually the similarity of model-based and census-based population trajectories (Kanyamibwa and Lebreton 1992). In this section, we provide a formal methodology for the simultaneous analysis of mark-recapture-recovery data and population information such as census data.

9.3.1 Data, Models and Integrated Modelling

We introduce integrated population modelling through an extensive example in which ring-recovery data from marked birds are combined with abundance data on the same species. However, the approach is quite general, as we shall see. Our example is described in Besbeas et al. (2002) and involves observations on lapwings *Vanellus vanellus* breeding in Britain. This is a species of conservation concern in the UK due to its dramatic decline in recent years and it has been placed onto

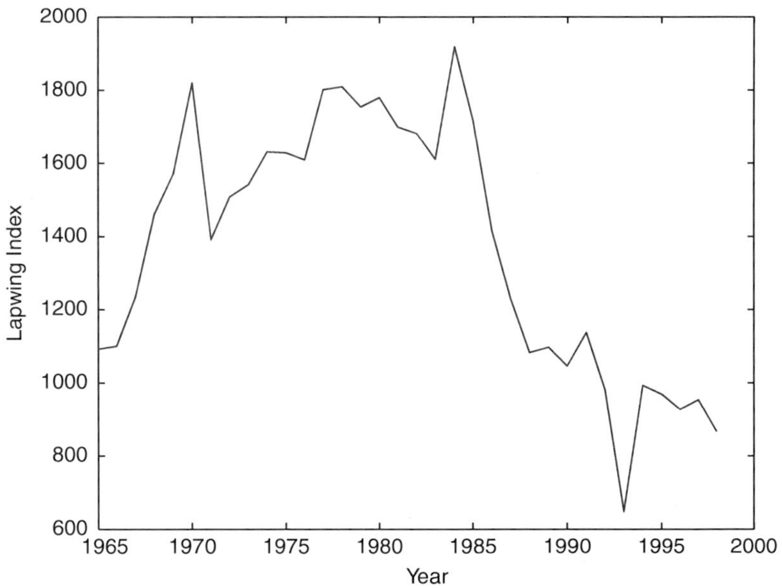

Fig. 9.1 The lapwing CBC index. The CBC index is sometimes given relative to a baseline year. As shown here, it is an estimate of the total population of lapwing territories for the set of sites included in the analysis

the Red List of species of highest conservation concern (Eaton et al. 2009). The census data we use are an index, derived from the common birds census (CBC) (Marchant et al. 1990). The CBC data are collected from specific survey sites by volunteers, and contain a large number of missing values. An index is constructed using a generalized linear model (ter Braak et al. 1994) which estimates site effects s_i (for site $i = 1, \ldots, S$) and year effects u_t (for year $t = 1, \ldots, T$) subject to an arbitrary constraint, e.g. $s_S = 0$. Annual index values y_t are then calculated as

$$y_t = \sum_{i=1}^{S} \exp(s_i + u_t), \quad t = 1, \ldots, T. \tag{9.1}$$

The resulting index estimates the relative abundance of the national breeding population from 1965 to 1998 inclusive, and it is plotted in Fig. 9.1.

The corresponding ring-recovery data provide the numbers of birds recovered dead in successive years after being ringed as chicks from 1963 to 1997. Note that these are national figures and are unlikely to share common individuals with the CBC data. The raw data are given in Besbeas et al. (2002), and as an illustration a subset of the recovery data is presented in Table 9.1.

We model the ring-recovery data using annual survival probabilities $\boldsymbol{\phi}$, with components which describe age-dependence, and a recovery probability κ, which

Table 9.1 Illustrative recovery sub-table for lapwings, taken from Besbeas et al. (2002)

Year of ringing	Number ringed	Year of recovery 1991	1992	1993	1994	1995	1996	1997	1998	No. of days below freezing (w)
1990	4170	12	3	3	2	1	0	2	0	12
1991	4314		9	4	6	1	0	1	0	12
1992	3480			18	3	1	2	0	1	9
1993	3689				6	5	2	2	1	6
1994	3922					12	4	6	0	3
1995	3591						7	5	1	18
1996	4488							7	0	10
1997	4339								5	0

The data show the numbers of British lapwings recovered dead in successive years after being ringed as nestlings, for birds ringed between 1990 and 1997 inclusive. Reporting probabilities of dead birds have been declining over time. By the 1990s, recovery rate had roughly halved relative to the 1960s. Also shown in the table is the number (w) of days in the year that a measure of Central England temperature fell below freezing. Note that year i ($1990 \leq i \leq 1997$) indicates the period of time from 1 April of year i to 31 March of year $(i + 1)$

Table 9.2 Cell probabilities $\{p_{ij}\}$ for a simple model for ring-recovery data, with no over-dispersion or time-variation.

Year of ringing	Year of recovery			
	1	2	3	4
1	$(1-\phi_1)\kappa$	$\phi_1(1-\phi_a)\kappa$	$\phi_1\phi_a(1-\phi_a)\kappa$	$\phi_1\phi_a^2(1-\phi_a)\kappa$
2		$(1-\phi_1)\kappa$	$\phi_1(1-\phi_a)\kappa$	$\phi_1\phi_a(1-\phi_a)\kappa$
3			$(1-\phi_1)\kappa$	$\phi_1(1-\phi_a)\kappa$

In this illustrative example, there are 3 years of ringing, 4 years of recovery, and three model parameters

is the probability of recovery and reporting of marked dead birds, and which varies over time. For the lapwings, there are two age classes of survival, corresponding to birds in their first year of life and older birds. Both of the annual survival probabilities are regressed on a single measure of winter severity w, using logistic regression. Thus we have $\text{logit}(\phi_1) = \beta_0 + \beta w$, for birds in their first year of life, with first-year annual survival probability of ϕ_1, and $\text{logit}(\phi_a) = \delta_0 + \delta w$, which applies to birds aged ≥ 1, with annual survival probability ϕ_a. In addition, the reporting probabilities for dead wild birds in Britain are generally found to be declining over time (Baillie and Green 1987), and so we set $\text{logit}(\kappa) = \nu_0 + \nu t$, where t measures year. See also McCrea et al. (2012b). We do not here consider over-dispersion, but if necessary that may be easily incorporated, for example by means of suitable additive random effects, as described by Barry et al. (2003), or through the use of beta distributions.

Morgan and Freeman (1989) and Freeman and Morgan (1992) describe more general models for ring-recovery data, involving additional age-dependence in survival, and/or time dependence in all parameters. See also Chap. 7. In order to write down the likelihood function, we introduce additional notation as follows. Let the number of birds ringed in year i be R_i, let the number recovered in year j, having been ringed in year i, be m_{ij}, and let the number which are not recovered from the year i cohort be $u_i = R_i - \sum_{j=i}^{c} m_{ij}$. Let the probability of recovery in year j given a bird was ringed in year i, corresponding to m_{ij}, be p_{ij} and let $q_i = 1 - \sum_{j=i}^{c} p_{ij}$ be the probability of non-recovery from the ith cohort. The recovery probability p_{ij} is a product of annual survival probabilities (from year i to year $j-1$, mortality in year j, and recovery in year j (e.g. Table 9.2). A particular model for the data consists of a specification of the probabilities $p_{ij} \equiv p_{ij}(\boldsymbol{\phi}, \kappa)$ in terms of the model parameters. In order to display appropriate multinomial cell probabilities, we shall take a ring-recovery study with birds ringed as nestlings for $r = 3$ successive years, and recoveries recorded for the $c = 4$ years following the initial ringing. In the simplest case, these parameters are constant, and the recovery probabilities are given in Table 9.2. For each cohort, the probabilities of non-recovery are (1− the corresponding row totals). Then, provided the birds suffer independent fates, both within and between cohorts, the likelihood for the ring-recovery data is product-multinomial in form, with log-likelihood given below. Here and later in this chapter we shall for convenience suppress the dependence of likelihoods on data.

$$\log L_r(\boldsymbol{\phi}, \kappa) = \sum_{i=1}^{r} \sum_{j=i}^{c} m_{ij} \log p_{ij} + \sum_{i=1}^{r} u_i \log q_i, \tag{9.2}$$

where terms not depending on the parameters have been omitted. We shall refer to Eq. (9.2) as the ring-recovery log-likelihood below.

The census data are described by means of a state-space model involving a productivity measure p and measurement error variance σ^2, in addition to the survival probabilities. For the lapwings we set $\log(p) = \gamma_0 + \gamma t$, to allow for a decline in productivity; Besbeas et al. (2005) consider instead a change-point in productivity, corresponding to the start of the decline in numbers. The elements involved in forming the census likelihood are as follows.

Let N_{1t} and N_{at} denote, respectively, the underlying numbers of one-year-old female birds and female birds aged ≥ 2 years at time t, and define $\mathbf{N}_t = (N_{1t}, N_{at})'$ and $N_t = N_{1t} + N_{at}$. If we assume no sex effect on survival and that breeding starts at age 2, then a natural process model for the underlying population sizes would be

$$N_{1,t+1} \mid \mathbf{N}_t \sim \text{Poisson}(N_{at}\, p\phi_1)$$

and

$$N_{a,t+1} \mid \mathbf{N}_t \sim \text{binomial}(N_t, \phi_a)$$

where the parameter p now denotes the annual productivity of females per female. The random variables N_{1t} and N_{at} can be approximated by appropriate independent normal variables, resulting in the following model

$$\begin{bmatrix} N_{1,t+1} \\ N_{a,t+1} \end{bmatrix} = \begin{bmatrix} 0 & p\phi_1 \\ \phi_a & \phi_a \end{bmatrix} \begin{bmatrix} N_{1t} \\ N_{at} \end{bmatrix} + \begin{bmatrix} \delta_{1t} \\ \delta_{at} \end{bmatrix}, \tag{9.3}$$

where the δ terms have zero means, and variances which are given by suitable Poisson and binomial expressions:

$$\begin{aligned} \text{Var}(\delta_{1t}) &= E(N_{a,t})p\phi_1 \\ \text{Var}(\delta_{at}) &= E(N_t)\phi_a(1 - \phi_a). \end{aligned}$$

If one is using classical inference, then it is necessary to use expectations in the variance expressions in order to comply with the assumptions of the Kalman filter; see Sullivan (1992) as well as the gray whale example in Sect. 6.4.2. The matrix above is a familiar Leslie matrix. Buckland et al. (2004, 2007) provide a general framework for deriving population projection matrices by considering intermediate sub-processes (Chap. 2). If we assume that only breeding birds are censused, then what we observe, y_t, which we take here as the CBC index, is given by the measurement equation,

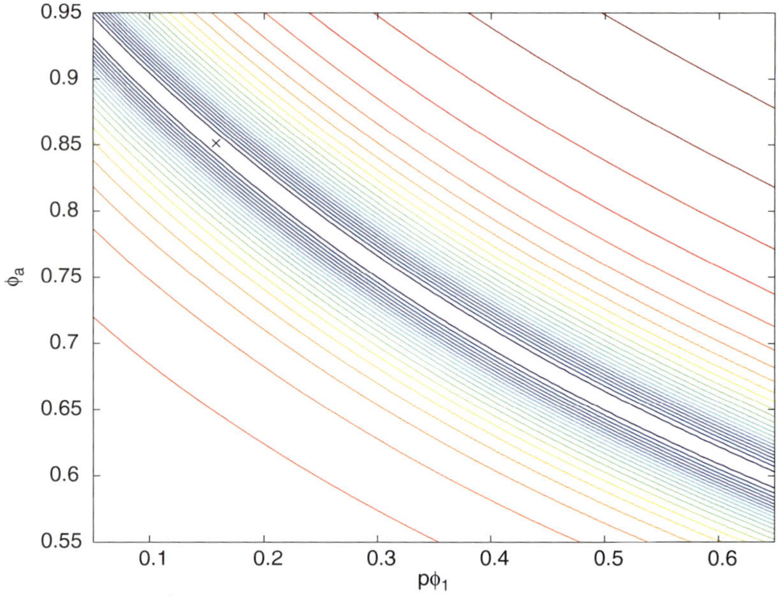

Fig. 9.2 Profile log-likelihood contours for $\theta = p\phi_1$ and ϕ_a from the lapwing census data, obtained by maximizing $\log L_c(p, \phi_1, \phi_a, \sigma)$ with respect to σ. The location of the maximum is shown by ×

$$y_t = (0 \;\; 1)\begin{pmatrix} N_{1t} \\ N_{at} \end{pmatrix} + \epsilon_t. \tag{9.4}$$

We assume that ϵ_t is normally distributed with constant variance σ^2, so that $\epsilon_t \sim \text{normal}(0, \sigma^2)$. (In some cases it may be sensible for σ^2 to vary with time.)

Equations (9.3) and (9.4) collectively form a normal dynamic linear model (see Sect. 4.4). Thus the likelihood for the census data, $L_c(p, \phi_1, \phi_a, \sigma)$, can be derived using the Kalman filter. However, the parameters p and ϕ_1 in this likelihood only appear as a product. Furthermore likelihood functions for census data generally provide limited information on the underlying demographic mechanisms individually. Figure 9.2 plots the two-dimensional profile log-likelihood contours of L_c for parameters $\theta = p\phi_1$ and ϕ_a for the lapwing data and illustrates how θ and ϕ_a are negatively correlated.

However ϕ_1 also occurs in L_r, and so we may obtain a full-rank model (see Sect. 5.2.2), in which all parameters may in principle be estimated, by maximizing the joint likelihood,

$$L_j(\phi_1, \phi_a, \kappa, p, \sigma) = L_r(\phi_1, \phi_a, \kappa) \times L_c(p, \phi_1, \phi_a, \sigma). \tag{9.5}$$

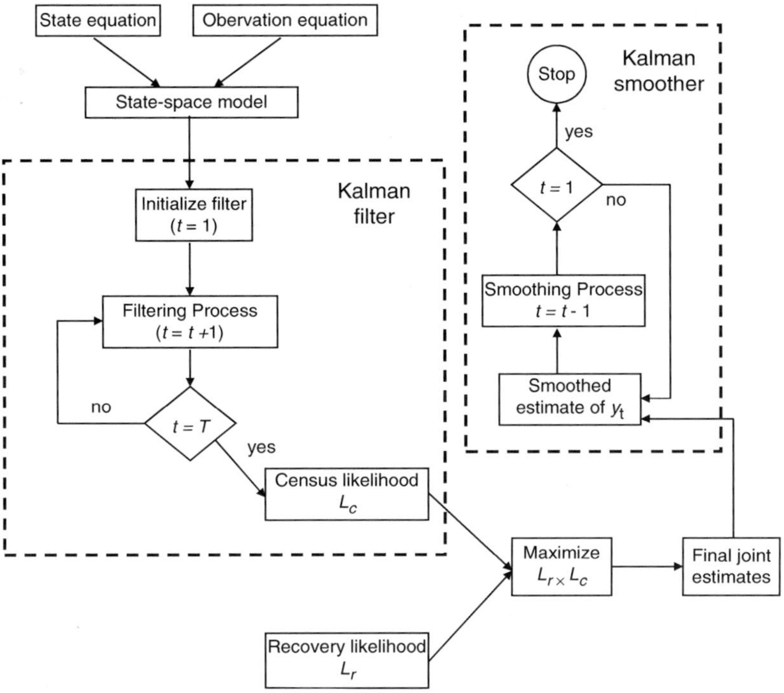

Fig. 9.3 Flow chart summarising the steps in fitting an integrated population model using classical inference incorporating the Kalman filter

(To reflect the various logistic/logarithmic regressions in the model, we can also write the joint likelihood as $L_j(\beta_0, \beta, \delta_0, \delta, \nu_0, \nu, \gamma_0, \gamma, \sigma)$, but we use the notation in Eq. (9.5) in a generic fashion, for brevity.) This is very useful, as the productivity of a species in decline is a parameter that we particularly want to investigate. The assumption of independence made in multiplying the two likelihoods together in Eq. (9.5) is not likely to be violated. The approach extends flexibly to include additional likelihood components, corresponding to data on say productivity or movement. This methodology has been termed *integrated population modelling* in the literature. The flow chart in Fig. 9.3 summarizes the steps of integrated population modelling, when classical inference is being used incorporating the Kalman filter. We shall now illustrate the performance of the integrated approach by application to the lapwing data before discussing a number of technical issues and extensions.

Table 9.3 Maximum likelihood parameter estimates from fitting the model $\phi_1(w), \phi_a(w)/\kappa(year)/p(year)$ to the lapwing data, (i) using ring-recovery data alone, and (ii) using integrated population modelling, incorporating both the ring-recovery data and the census data

		Parameter estimates		Estimated standard errors	
		(i)	(ii)	(i)	(ii)
ϕ_1	Intercept (β_0)	0.5158	0.5231	0.0675	0.0672
	Slope (β)	−0.0241	−0.0228	0.0072	0.0070
ϕ_a	Intercept (δ_0)	1.5011	1.5210	0.0683	0.0690
	Slope (δ)	−0.0360	−0.0279	0.0051	0.0045
κ	Intercept (ν_0)	−4.5668	−4.5632	0.0350	0.0350
	Slope (ν)	−0.5729	−0.5841	0.0641	0.0636
p	Intercept (γ_0)		−1.1513		0.0880
	Slope (γ)		−0.4323		0.0743
σ			159.4691		21.8712

The estimated standard errors result from inverting a numerical approximation to the Hessian matrix at the maximum likelihood estimate. Reprinted with permission from *Biometrics*

9.3.2 Results for Lapwing Example

The decline of lapwings may be described through inclusion of time in survival and/or productivity parameters. Here we follow Besbeas et al. (2002) and allow the productivity parameter p to be a function of time: $p(year)$. Thus we have fitted to the lapwing data the model,

$$\phi_1(w), \phi_a(w)/\kappa(year)/p(year),$$

which indicates that both ϕ_1 and ϕ_a are logistic functions of the single covariate, w, which measures the number of days in the year when the temperature at a location in central England was below freezing. We concentrate on logistic regressions, as described in the last section, but alternative link functions might also be used. Additionally, we let $\kappa(year)$, $p(year)$ denote respectively logistic and logarithmic regressions of κ and p on year. Note that p is not a probability, and so is not bounded above by unity. The maximum likelihood point estimates from the joint data, and also from the recovery data only, are given in Table 9.3.

We can see that the joint analysis changes slightly the maximum likelihood estimates from the ring-recovery data, as now the estimates describe both the data sets. There is little difference between the standard errors of the common parameters in the two analyses, due to the dominance of the ring-recovery data in this illustration. The same conclusion arises from a Bayesian analysis; see Brooks et al. (2004). When we repeated the analysis with a subset of the recovery data, then we found that the joint analysis produced substantially reduced estimates of

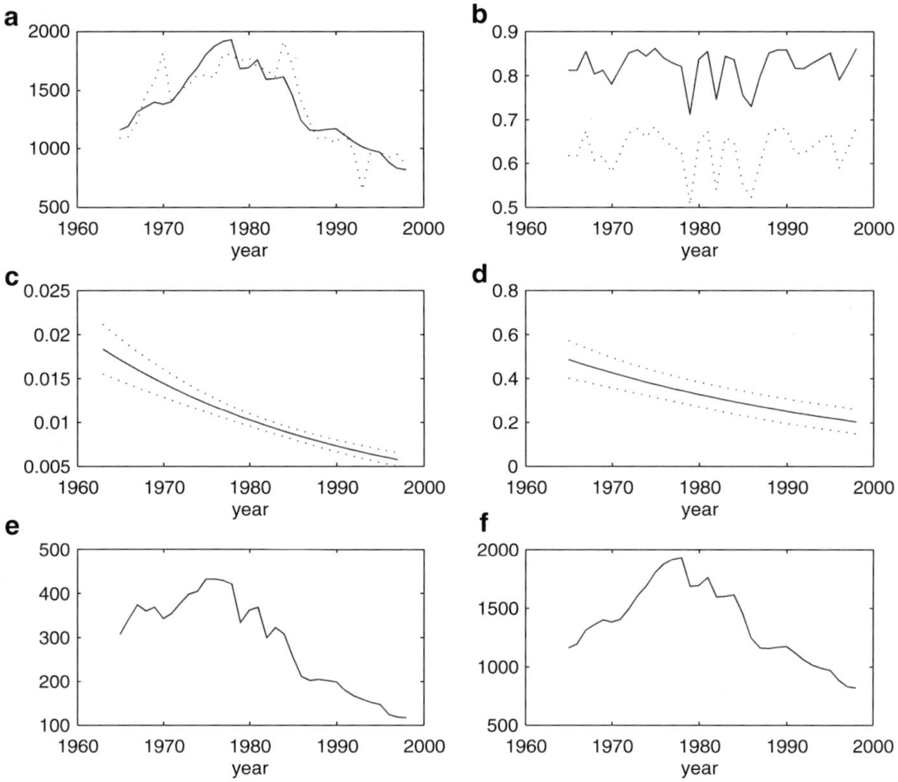

Fig. 9.4 The result of fitting model $\phi_1(w), \phi_a(w)/\kappa(year)/p(year)$ to the lapwing data. (**a**) Observed census data (*dotted line*); fitted curve (*solid line*). (**b**) $\hat{\phi}_a$ (*solid line*); $\hat{\phi}_1$ (*dotted line*). (**c**) $\hat{\kappa}(year)$, with estimated 95 % confidence bands. (**d**) $\hat{p}(year)$, with estimated 95 % confidence bands. (**e**) Graph of $\hat{N}_{1t}$; (**f**) Graph of $\hat{N}_{at}$, equivalent to the fitted curve in (**a**). Reprinted with permission from *Biometrics*

standard error compared with the ring-recovery ones; see Besbeas et al. (2002). For the ring-recovery data alone, due to the length of the study, correlations between parameter estimators are typically quite small, ranging in magnitude from −0.20 to 0.29. The addition of the census data has little effect on those correlations. However there are now correlations of −0.49 between $\hat{\gamma}_0$ and $\hat{\beta}_0$, and −0.91 between $\hat{\gamma}_0$ and $\hat{\delta}_0$, where γ_0, β_0 and δ_0 are defined in Table 9.3. These are sensible findings, since increasing the productivity p requires a decrease in survival in order to match the data. We would also expect a stronger correlation with the intercept estimate of the adult survival, $\hat{\delta}_0$, as is seen above.

We show in Fig. 9.4 the results from maximizing the combined likelihood L_j. The confidence bands shown result from applying the δ-method. The decline of κ with time agrees with Catchpole et al. (1999). However of greater interest to us here is the time-varying behaviour of the parameters ϕ_1, ϕ_a and p. We can see

Table 9.4 The results from fitting by maximum likelihood a range of models to the lapwing data

(i)

Model	$-\ell$	No. of parameters	AIC
$\phi_1(w), \phi_a(w)/\kappa(year)$	7156.33	6	14325
$\phi_1(w, year), \phi_a(w, year)/\kappa(year)$	7155.16	8	14326

(ii)

Model	$-\ell$	No. of parameters	AIC	ΔAIC
$\phi_1(year), \phi_a(year)/\kappa(year)/p$	7409.54	8	14835	52
$\phi_1(w), \phi_a(w)/\kappa(year)/p$	7398.40	8	14813	30
$\phi_1(w), \phi_a(w, year)/\kappa(year)/p$	7382.37	9	**14783**	0
$\phi_1(w, year), \phi_a(year)/\kappa(year)/p$	7401.14	9	14820	37
$\phi_1(w, year), \phi_a(w, year)/\kappa(year)/p$	7381.45	10	**14783**	0
$\phi_1(year), \phi_a(year)/\kappa(year)/p(year)$	7409.29	9	14837	54
$\phi_1(w), \phi_a(w)/\kappa(year)/p(year)$	7383.38	9	**14785**	2
$\phi_1(w), \phi_a(w, year)/\kappa(year)/p(year)$	7381.99	10	**14784**	1
$\phi_1(w, year), \phi_a(w)/\kappa(year)/p(year)$	7383.36	10	14787	3
$\phi_1(w, year), \phi_a(w, year)/\kappa(year)/p(year)$	7381.45	11	14786	3

In (i) we fit only the ring-recovery data; in (ii) we use integrated population modelling to fit a range of alternative models to both ring-recovery and census data. Here ℓ denotes the value of the log-likelihood evaluated at the maximum likelihood estimates of the parameters. AIC denotes the Akaike Information Criterion, and ΔAIC indicates the difference between the model AIC value and the smallest AIC value for the set of models considered. In (ii) we show in bold type the AIC values corresponding to alternative acceptable models for the combined data set, with $\Delta AIC \leq 2$. Reprinted with permission from *Biometrics*

that the decline in lapwing numbers since 1980 is compatible with a major drop in the productivity parameter p. An alternative explanation of the recent decline in lapwing numbers is that there is a decline in survival probability over time. There was no evidence for this in Catchpole et al. (1999), or for the more extensive ring-recovery data set analysed here (see Table 9.4(i)).

However we can see from Table 9.4(ii) that when we analyse the combined data using integrated population modelling, then in terms of AIC, several models provide comparable best fit to the data, in particular with constant p and declining probabilities of survival over time. Detailed studies of breeding lapwings have shown a decrease in chicks produced over the period. This is usually attributed to the switch from spring to autumn sowing of cereals and intensification of pasture management (see Wilson et al. 2001 and references therein). Thus in order to demonstrate model performance, we shall here only consider the model $\phi_1(w), \phi_a(w)/\kappa(year)/p(year)$, as in Besbeas et al. (2002). For further discussion, see King et al. (2008).

We show in Fig. 9.5 the regressions of $\hat{\phi}_1$ and $\hat{\phi}_a$ on w, combined with plots of $\hat{\phi}_{1,t}$ and $\hat{\phi}_{a,t}$ resulting from a model with separate ϕ_1 and ϕ_a parameters for each year, and denoted by $\{\phi_{1,t}\}$ and $\{\phi_{a,t}\}$ respectively. These regressions are seen to provide a fair description of the relationship between annual survival and w. Note that the

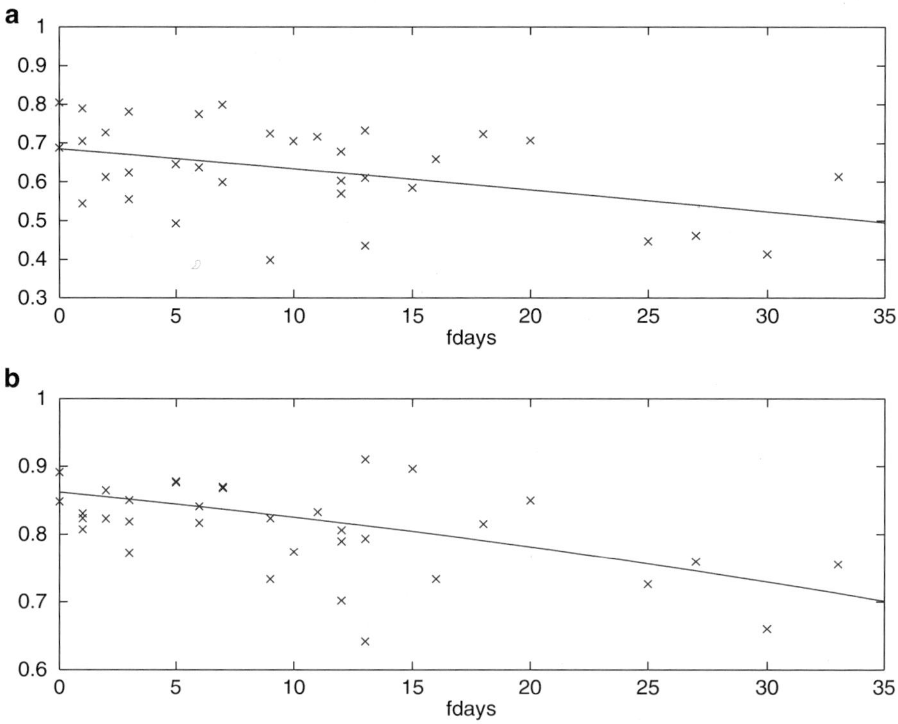

Fig. 9.5 A graphical demonstration of the logistic regressions of ϕ_1 and ϕ_a in the model, $\phi_1(w), \phi_a(w)/\kappa(year)/p(year)$, for the lapwing data. Here the covariate w denotes the number of days in a year when the temperature in central England was below freezing. In (**a**) we graph $\hat{\phi}_1$ against w and in (**b**) we graph $\hat{\phi}_a$ against w. Also plotted are the values $\{\hat{\phi}_{1t}\}$ and $\{\hat{\phi}_{at}\}$ respectively, corresponding to having a separate value for ϕ_1 and ϕ_a for each year. Reprinted with permission from *Biometrics*

estimates of β and δ in Table 9.3 are approximately equal, and a model in which ϕ_1 and ϕ_a share a common slope parameter produces virtually no change to the likelihood.

Shown in Fig. 9.4e is the smoothed estimate of N_{1t}. The figure demonstrates the large decline in N_{1t} in recent years, in line with the predicted decline in p over time in this model.

We do not provide a formal test of goodness-of-fit of the selected model, but in Fig. 9.6 we plot the observed numbers of recoveries against the fitted numbers, and also provide a Q-Q plot of the prediction errors from the Kalman filter, which are expected to have a normal distribution. In the latter case, the single large prediction error is due to the initial census value, and is to be expected because of the way in which the Kalman filter analysis is initiated. Overall there is no serious indication of systematic lack of fit.

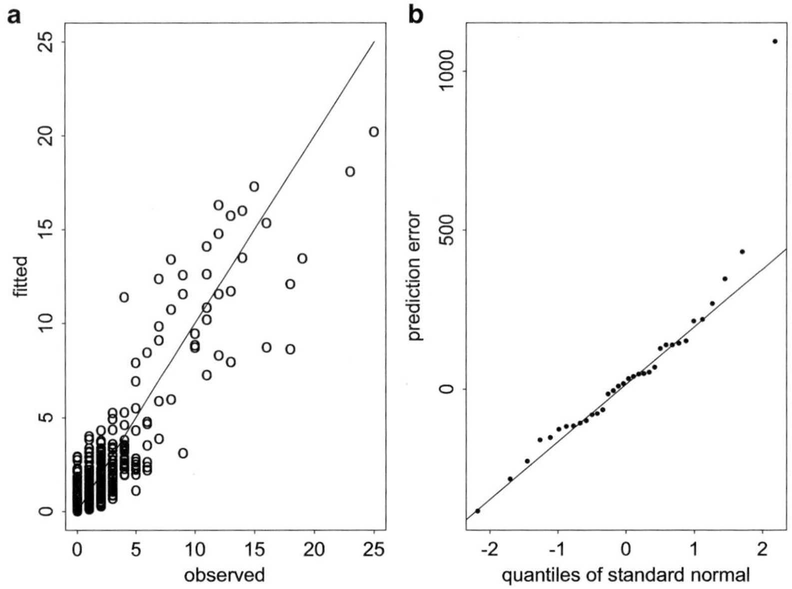

Fig. 9.6 Graphical checks of goodness-of-fit of the model $\phi_1(w), \phi_a(w)/\kappa(year)/p(year)$, for the lapwing data: (**a**) observed numbers plotted vs expected numbers for the recovery data; (**b**) a Q-Q plot of the prediction errors from the Kalman filter. Reprinted with permission from *Biometrics*

9.4 Facilitating Connected Likelihood Modelling

9.4.1 Normal Approximation

The integrated approach described above requires specialized computer code, not only for the Kalman filter component, but also for the additional likelihood components, some of which may be complex, and may even have been derived using specialist computer packages. Quite often, for example, survival data would be analysed by Program MARK (White and Burnham 1999). This then poses obvious difficulties for combined analysis and is likely to preclude the use of the integrated approach in practice. A solution to this problem is provided by an approximation, suggested and evaluated by Besbeas et al. (2003). Here, a multivariate normal approximation is adopted for the form of the likelihood of the ring-recovery data, making use of the parameter estimates, and their corresponding estimates of dispersion obtained from analysing the ring-recovery data alone. In particular, we make a multivariate normal approximation to $L_r(\boldsymbol{\phi}, \kappa)$:

$$\begin{aligned} \log L_r(\boldsymbol{\phi}, \kappa) &\approx \text{constant} - \frac{1}{2}(\hat{\boldsymbol{\theta}} - \boldsymbol{\theta})'\hat{\boldsymbol{\Sigma}}^{-1}(\hat{\boldsymbol{\theta}} - \boldsymbol{\theta}) \\ &\equiv \log \hat{L}_r(\boldsymbol{\phi}, \kappa) \end{aligned} \tag{9.6}$$

where we write $\boldsymbol{\theta}$ to denote the model parameters on the logistic scale, and where $\hat{\boldsymbol{\theta}}$ and $\hat{\boldsymbol{\Sigma}}$ are respectively the maximum likelihood estimates of $\boldsymbol{\theta}$ and the dispersion matrix of $\hat{\boldsymbol{\theta}}$, both obtained from a separate model-fitting exercise for the ring-recovery data alone. The approximation is motivated naturally by the asymptotically multivariate normal distribution of maximum likelihood estimators, and has been used to good effect in Lebreton et al. (1992). The integrated population modelling then proceeds by replacing the exact likelihood $L_r(\boldsymbol{\phi}, \kappa)$ in (9.5) with expression (9.6):

$$L_j(\phi_1, \phi_a, \kappa, p, \sigma) = \hat{L}_r(\phi_1, \phi_a, \kappa) \times L_c(p, \phi_1, \phi_a, \sigma).$$

The approach has great potential to approximate likelihood components for abundance data obtained from complex sampling schemes (Knape et al. 2013). This means that particular programs or packages for survival and also for fecundity or other data, if appropriate, need only be run once, to obtain the relevant maximum likelihood estimates of relevant parameters and their estimates of standard error and correlation. This clearly greatly simplifies both the resulting form for L_j and its maximization.

9.4.2 Results

For the example of the last section, we evaluate the approximation in Table 9.5; the standard errors are obtained from the observed information matrices as in Table 9.3. The agreement between exact and approximate results is seen to be very good. An additional benefit from using the approximate approach is that in comparison with the exact analysis, it is far less sensitive to starting values for the maximum likelihood iteration. Additionally the approximate analysis was found to be about 2.5 times faster than the exact method.

We can see from Table 9.5 that the change in the value of $\hat{\boldsymbol{\theta}}$ from the recovery analysis alone to the joint analysis is not large. It is both the overall magnitude of this change and the size of the data sets which determine the effectiveness of the multivariate normal approximation made in the paper, as we require the multivariate normal approximation to L_r to be good for the value of $\boldsymbol{\theta}$ which maximizes L_j, and not just for the value that maximizes L_r. We show in Fig. 9.7 the good agreement between exact and approximate $\log L_r(\boldsymbol{\phi}, \kappa)$ for the small illustrative data set of Table 7.1, by means of profile log-likelihoods, for the parameter β_0. We have found the good agreement observed in Fig. 9.7 also for profiles with respect to other parameters. Besbeas et al. (2003) provide further details. The good performance of the approximation extends to the case of multi-site data (McCrea et al. 2010).

Table 9.5 An evaluation of the multivariate normal approximation for the ring-recovery likelihood for the lapwing data

Parameter	Ring-recovery alone	Exact combination	Approximate combination
β_0	0.5158 (0.0679)	0.5231 (0.0679)	0.5226 (0.0678)
β	−0.0241 (0.0072)	−0.0228 (0.0070)	−0.0227 (0.0070)
δ_0	1.5011 (0.0685)	1.5210 (0.0693)	1.5191 (0.0686)
δ	−0.0360 (0.0051)	−0.0279 (0.0045)	−0.0280 (0.0045)
ν_0	−4.5668 (0.0351)	−4.5632 (0.0352)	−4.5634 (0.0351)
ν	−0.5729 (0.0640)	−0.5841 (0.0637)	−0.5837 (0.0638)
γ_0		−1.1513 (0.0886)	−1.1489 (0.0876)
γ		−0.4323 (0.0743)	−0.4314 (0.0740)
σ		159.469 (22.062)	159.613 (21.875)

In each case we show the maximum likelihood parameter estimate and the corresponding standard error. The estimates of error are obtained as in Table 9.3. Reproduced with permission from *Applied Statistics*

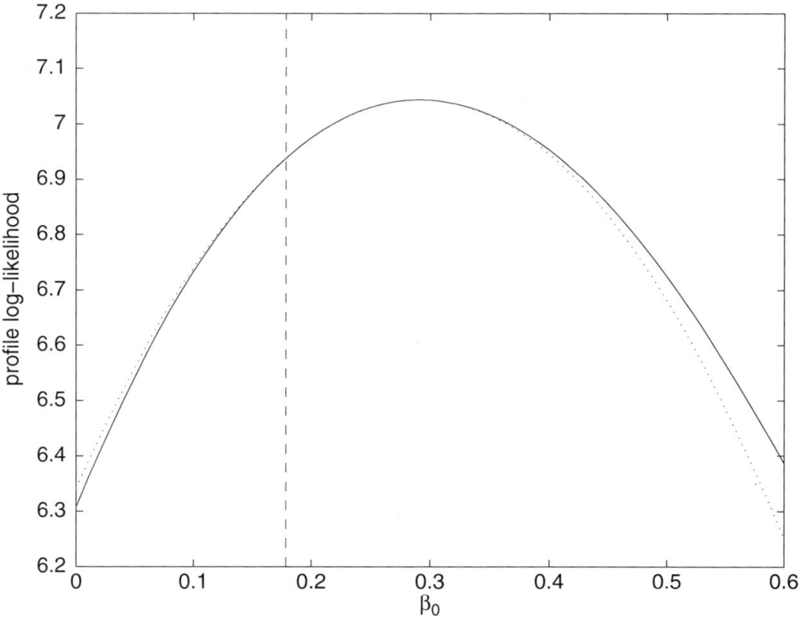

Fig. 9.7 Agreement between $\log L_r(\boldsymbol{\phi}, \kappa)$ and the multivariate normal approximation for the data of Table 7.1. The two curves are profile log-likelihoods taken with respect to the parameter β_0. The *solid curve* is the exact likelihood, and the *dotted curve* approximate. Also shown is the location of $\hat{\beta}_0 = 0.1786$, the value which maximizes the combined exact likelihood, making use of the entire run of the census data for lapwings. Reprinted with permission from *Applied Statistics*

9.5 Technical Issues for Classical Analysis

The connected likelihood approach depends on making a number of assumptions and adopting certain procedures. For example, we assume that we can approximate discrete distributions by normal distributions; that we can suitably start the Kalman filter iterations; that the different surveys are independent; and that the state-space model adopted correctly partitions variation between its transition and measurement processes. Brooks et al. (2004) demonstrated the robustness of the normality assumptions. In this section, we explore the other issues listed above, starting with the problem of initializing the Kalman filter.

9.5.1 Kalman Filter Initialization

Computational algorithms in state space analyses are mainly based on recursions in which we calculate values at time t from earlier values for $t-1,\ldots,1$. The question of how these recursions are started up at the beginning of the series is called initialization. The initialization problem of the Kalman filter when $\boldsymbol{\mu}_0$ and $\mathbf{Q}_0$ are unknown is an issue requiring attention in general, in areas such as economics and engineering, but the problem may be more important in population ecology, where there are often small samples, models may involve a large number of states and unknown parameters, and models are usually non-stationary. We shall now describe an approach to initialize the Kalman filter which is designed for integrated population modelling in ecology.

9.5.1.1 Stable-Age Initialization

In population ecology, the state vector $\mathbf{n}_t$ is typically a vector denoting the numbers of individuals in the population in a number of classes at time t. Typical elements of $\mathbf{A}_t$ include age- or stage-specific survival and productivity parameters and rates at which individuals in one state make the transition to another state, for example through immigration or emigration. The matrix $\mathbf{A}_t$ is referred to as a Leslie or Lefkovitch matrix, depending on whether the population is age- or stage-classified, respectively (Chap. 2). We encountered a Leslie matrix in Sect. 9.3.1.

The Perron-Frobenius theorem states that, for appropriate *constant* transition matrices $\mathbf{A}$, there exists a real positive eigenvalue κ that is greater in absolute value (or in modulus, if some of the other eigenvalues are complex) than all of the other eigenvalues. The implications are that the dominant eigenvalue κ represents the asymptotic growth rate of the population, and the normalized right eigenvector associated with κ represents the asymptotic proportion of every age or stage class in the total population. We call the eigenvalue κ *the asymptotic growth rate* and its corresponding right eigenvector, $\mathbf{v}$, is called the *stable age* (or *stable stage*) distribution.

We propose starting the Kalman filter by taking the initial mean vector $\boldsymbol{\mu}_0$ to be proportional to the stable age (stage) distribution of a Leslie (Lefkovitch) matrix **A**, with the proportions scaled by the total size of the first observation, $\mathbf{n}_0$. To choose $\mathbf{Q}_0$, we adopt the conservative approach of requiring the lower end of an appropriate confidence interval for each element of $\mathbf{n}_0$ to be non-negative, and that elements are independent. Thus for example for univariate observations and 95 % confidence intervals we take

$$\boldsymbol{\mu}_0 = \mathbf{v}y_1/(\mathbf{Z}_1\mathbf{v}), \qquad \mathbf{P}_1 = \text{diag}((\boldsymbol{\mu}_0/1.96)^2)$$

where y_1 and $\mathbf{Z}_1$ are the first observation and measurement vector respectively. In practice, in order to derive the stable age distribution, we need to know **A**, which may contain unknown parameters. We select the values for the parameters in **A** that are in common with the demographic analyses by using their maximum likelihood estimates obtained from analysing the demographic data alone. Any remaining parameter(s) can then be obtained by iteration; for details see Besbeas and Morgan (2012b). For instance if productivity p is unknown, then we take an arbitrary value for p, and obtain a maximum likelihood fit, which provides an estimate, $\hat{p}^{(1)}$. We then use this value to start the Kalman filter, resulting in an estimate $\hat{p}^{(2)}$, and so on. When the matrix $\mathbf{A}_t$ is time-dependent, then we obtain **A** by an appropriate time-average of the $\mathbf{A}_t$. The good performance of this approach compared with alternatives in which the elements of the initial state vector are diffuse, that is, treated as random variables with infinite variance, or are treated as unknown constants to be estimated by maximum likelihood, is shown in simulation studies presented by Besbeas and Morgan (2012b).

9.5.2 *Lack of Independence*

It may be that census data and demographic data are not completely independent. Besbeas et al. (2009) consider the effect of dependence between a ring-recovery data set and census data. This was done for a model with two age-classes for survival, as in Sect. 9.3.1, and with constant parameters, ϕ_1, ϕ_a and κ. Life histories spanning eight years were constructed for a 20-year period, with $\phi_1 = 0.5$, $\phi_a = 0.7$, for a range of values of κ, and probability of recapture of a live animal. Observation error was added to the life histories, resulting in census data. It was shown that, in some circumstances, combining dependent data sets but treating them as independent can actually reduce estimator precision. While this was only a single study, the message is that one should take care conducting combined analyses for dependent data sets.

9.5.3 *Heterogeneity in the State-Space Model*

In Sect. 9.3.1, we used binomial error variances for when a number n of individuals survive or die in any particular year, and Poisson error variances for recruitment arising from reproduction. In both of these cases, we can allow for heterogeneity. A study with an application to data on grey herons *Ardea cinerea* is provided by Besbeas et al. (2009) in which census data are combined with ring-recovery data. It was found that when overdispersion is included in the integrated modelling, it is possible to obtain $\hat{\sigma} = 0$. We return to this finding in the next section. Thus while it is in principle straightforward to include overdispersion in integrated population modelling, there may be an interaction between the roles of the overdispersion parameter and the parameter denoting measurement error.

9.6 From Modelling an Index to Multi-Sites: Additional Complexity

In this section we consider how the connected likelihood approach can be extended to multi-site and multi-state models.

9.6.1 *Accounting for Different Habitats*

The CBC sites on which the index y_t was based in Sect. 9.3.1 may be classified as arable, grazing, mixed (i.e. arable and grazing) or "other" (which are not farmland and could include estuaries, for example). The study by Besbeas et al. (2002) did not make use of this information but it is interesting to consider how it might be used to give a breakdown with respect to habitat. For illustration, we outline the integrated analysis of the grazing and arable sites, which account for 31 % of the sites; Besbeas et al. (2005) give detailed results and also discuss the complete set of results from all four habitats.

The ring-recovery component estimates overall survival, and does not produce a breakdown with respect to habitat. However we use the state-space model given below, using superscripts A and G to indicate arable and grazing, respectively. Let $\mathbf{n}_t = (n_{1t}^A, n_{at}^A, n_{1t}^G, n_{at}^G)'$ and $\mathbf{y}_t = (y_t^A, y_t^G)'$. Then, with a general notation for respective error terms $\boldsymbol{\delta}_t$ and $\boldsymbol{\epsilon}_t$,

$$\mathbf{n}_t = \mathbf{A}\mathbf{n}_{t-1} + \boldsymbol{\delta}_t, \qquad \mathbf{y}_t = \mathbf{B}\mathbf{n}_t + \boldsymbol{\epsilon}_t,$$

where

$$\mathbf{A} = \begin{bmatrix} 0 & p^A\phi_1 & 0 & 0 \\ \phi_a & \phi_a & 0 & 0 \\ 0 & 0 & 0 & p^G\phi_1 \\ 0 & 0 & \phi_a & \phi_a \end{bmatrix} \quad \text{and} \quad \mathbf{B} = \begin{bmatrix} 0\,1\,0\,0 \\ 0\,0\,0\,1 \end{bmatrix}.$$

Thus n^A_{1t}, for example, is the number of one-year-old female lapwings fledged, and assumed subsequently to live, on arable land at time t, and y^A_t denotes the CBC index value at time t, derived from the arable sites only. Thus the model now assumes different productivities for arable sites, p^A, and grazing sites, p^G, and we also assume that there are different observation error standard deviations, σ^A and σ^G for arable and grazing respectively. Note that the model analyses the arable and grazing populations separately and assumes no movement of individuals between the two habitats. We discuss extensions to multi-site modelling incorporating movement probabilities in Sect. 9.6.3. The likelihood function from this multivariate model is equivalent to the product of likelihood functions for univariate data

$$L_c(\boldsymbol{\phi}, p^A, \sigma^A; \mathbf{y}^A) \times L_c(\boldsymbol{\phi}, p^G, \sigma^G; \mathbf{y}^G)$$

where $\boldsymbol{\phi} = (\phi_1, \phi_a)$, and $\mathbf{y}^A$ ($\mathbf{y}^G$) denote all the observations on arable (grazing) sites over time. These likelihood functions are constructed for the relevant index alone, as in Sect. 9.3.1. The likelihood can readily be extended to deal with more than two indices as well as individual site data. The analysis then follows the same lines as above, based on the model

$$\phi_1(w), \phi_a(w), \kappa(year), p^A(year), p^G(year).$$

The results indicate that productivity varies with habitat; see Besbeas et al. (2005) for details.

9.6.2 Modelling Individual Site Data

Besbeas and Freeman (2006) provide an alternative approach to dealing with individual site data which sidesteps the intermediate process of deriving an index of abundance. The approach is single-stage and has several advantages, including no loss of information in summarizing the raw data using an index. This approach fits the survey data directly by incorporating the population model into the generalized linear model used to derive the index in the existing approach. For example, from Eq. (9.1) and the structural part of the state-space model of Eqs. (9.3) and (9.4), we obtain the recursive relationship for the year effects

$$u_t = \phi_a u_{t-1} + p\phi_1\phi_a u_{t-2}, \quad t > 2, \tag{9.7}$$

which gives rise to the likelihood function, $L_{glm}(\phi_1, \phi_a, p, u_{1:2}, \mathbf{s})$. The form of L_{glm} depends on the distributional assumptions made for the individual site data, c_{it}. For example, when the c_{it} are assumed to follow independent Poisson distributions, then the log-likelihood function is given by

$$\log L_{glm}(\phi_1, \phi_a, p, u_{1:2}, \mathbf{s}) = \text{constant} + \sum_{i=1}^{S}\sum_{t=1}^{T} c_{it}(s_i + u_t) + e^{s_i + u_t}$$

where $u_{1:2} = (u_1, u_2)$ and u_t is given by Eq. (9.7) for $t > 2$. The integrated population modelling procedure then replaces the component corresponding to the abundance index, $L_c(\phi_1, \phi_a, p, \sigma)$, in Eq. (9.5), by $L_{glm}(\phi_1, \phi_a, p, u_{1:2}, \mathbf{s})$. The procedure is shown to perform well relative to analysing an index; see also Maunder (2001) for related work. Freeman and Besbeas (2012) provide further development with respect to analysing presence-absence survey data. The ease with which the connected likelihood approach can integrate different types of information, assuming independence between the sources, makes it very appealing in practice.

9.6.3 *Multi-Site and Multi-State Modelling*

The work described so far assumes no movement of individuals. How the general theory of integrated population modelling extends to the multi-site case is described by McCrea et al. (2010). The illustrative example of that paper involved great cormorants *Phalacrocorax carbo sinensis*, moving between three sites in Denmark, with the additional complication that birds could move between breeding and non-breeding states, as well as between sites. Model selection was based on a step-up procedure applied to the recapture data alone, in order to avoid possible problems with overfitting the census data, as discussed in Sect. 9.5.3. The final integrated population model included complex state- and site-dependent transitions. There was a pronounced improvement in the precision of estimators from combining the two data sets in a single analysis.

9.7 Additional Aspects of IPM

9.7.1 *Conditional Gaussian Modelling*

The main advantage of analysing census data using normal dynamic linear models lies in the use of the Kalman filter, which greatly facilitates the estimation process. Despite the conceptual simplicity of this model, it is both flexible and general, and extends to a range of other models, such as linear matrix population models which have wide application in ecology and demography (Caswell 2001). The non-linear case however is important in ecological settings, for example in accommodating density dependence in the population model or allowing observation variance to depend on population size; see Sect. 4.4.2. In certain important cases, non-linear models are still amenable to analysis by the Kalman filter. These models are known as *conditional Gaussian models* and are described in detail in Harvey (1989, Sect. 3.7.1). Here system matrices $\mathbf{A}_t$, $\mathbf{B}_t$, $\mathbf{Q}_t$ and $\mathbf{R}_t$ may depend upon previous observations, up to and including $\mathbf{n}_t$ in the notation of Eq. (4.9). The essence of these models is that even though the matrices may depend on observations up to time t, they may be regarded as fixed once time t has been reached, and thus the derivation of the Kalman filter still applies.

Besbeas et al. (2009), Tavecchia et al. (2009) and Besbeas and Morgan (2012a) provide illustrations of this approach on three different species. The illustration below is taken from this last paper, based on the grey heron. This example also involves ring-recovery and census data but uses a more elaborate survival age structure than the lapwing illustration, involving four age classes. The heron census data have been published widely and a notable characteristic of this species, other than its marked population crashes, is that the population rebounds quickly after a crash.

Besbeas and Morgan (2012a) considered several models for heron productivity p_t in year t, including a direct density-dependent model

$$\log p_t = \gamma_0 + \gamma y_t,$$

and a threshold model, where for a suitable threshold, τ,

$$\log p_t = \begin{cases} \nu_0 + \nu_1 & \text{if } y_t < \tau, \\ \nu_0 & \text{if } y_t \geq \tau. \end{cases}$$

The motivating assumption of this model is that as the birds nest in heronries, then when numbers are low, there might be less competition for space and resources, resulting in higher productivity than when the numbers are high. We would thus expect $\nu_1 > 0$. The threshold model is found to perform well relative to alternative models; see Besbeas and Morgan (2012a) for details, including extensions to more than one threshold. This paper also compares the results of a conditional Gaussian model with those from a Bayesian approach.

9.7.2 Goodness of Fit

A global goodness-of-fit test does not exist for integrated population models. For IPMs using a single SSM, the methods discussed in Sect. 5.6, e.g. recursive residuals, are a possibility.

For connected likelihood models, it seems intuitively reasonable to examine goodness of fit for each component of the model separately, which enables us to use off-the-shelf goodness-of-fit techniques in each of MRR modelling and time-series modelling, for example. Thus, in the case of mark-recovery and census data, as in Sect. 9.3.1, it is possible to examine plots of observed versus fitted numbers of recoveries which will indicate the fit of the mark-recovery component of the model, whilst for the census component, the prediction errors $\mathbf{v}_t$ are expected to be normally distributed and thus it might be appropriate to examine a Q-Q plot of these errors. Figure 9.6 provides an example of this approach for the lapwing data. The per-component goodness of fit can also be examined using standard tests, such as chi-square tests or numerical tests for normality. The extension to other types of mark-recapture-recovery data is obvious. However in all cases, it is important to note that integrated population modelling estimates are joint estimates, describing several components simultaneously, and may thus inherently manifest some lack-of-fit to each component individually.

The time-series plot of the observed and corresponding smoothed population estimates is an additional, natural diagnostic tool for the census component to examine whether the fitted model exhibits the stylised characteristics concerning the series. This plot is illustrated in Fig. 9.4a for the lapwing data. The time-series plot of the smoothed states against time is also a valuable diagnostic tool to check if the components extracted provide a suitable representation of these characteristics, and these are illustrated in Fig. 9.4e and f. The estimated observation variance provides a quantitative indication of the overall fit, with smaller values indicating a better model fit. However as discussed in the last section, it is important to consider this approach in parallel with model selection and heterogeneity in the state-space model, and the approach may need to be informed by additional considerations, which might for instance provide guidance on the appropriate magnitude of observation error. The potential to over-fit census data exists as shown in Besbeas et al. (2009), and any integrated population modelling should conclude with a particular assessment of the estimated measurement error, and consideration of whether or not it is appropriate.

In recent work, Besbeas and Morgan (2014) illustrate how Monte Carlo simulation can be used in the evaluation of goodness of fit of integrated models. We expect goodness-of-fit Monte Carlo simulation techniques to be become increasingly adopted both in integrated modelling and in capture-recapture in the future.

9.7.3 Bayesian Methods

The emphasis in this chapter has been on the connected likelihood approach using methods of classical analysis. Bayesian methods for state-space modelling have been described in Sect. 4.3. See also, for example, Meyer and Millar (1999) and Millar and Meyer (2000a,b). These papers involve fisheries applications, where there have been numerous other applications— see for example Rivot et al. (2001,

2004), Rivot and Prévost (2002) and Rifflart et al. (2006). The Bayesian approach to integrated population modelling was initially described by Brooks et al. (2004) and for more information, as well as simple `WinBUGS` programs to perform the necessary Markov chain Monte Carlo iterations, see Schaub et al. (2007), Gimenez et al. (2009b) and King et al. (2009). A clear advantage of the Bayesian approach is that it removes the need to use the Kalman filter, and hence the need to make normal approximations, as well as to specify the variances of the δ terms of Eq. (9.3) in terms of expectations. However the work of Brooks et al. (2004) showed that classical methods are robust to the normal approximation. We can also model density dependence directly with a Bayesian approach, and avoid the conditionally Gaussian modelling of the last section. In addition, goodness of fit can be assessed using Bayesian p-values; see Sect. 5.6.4. However King (2011) discusses how MCMC can be slow for fitting SSMs. The interplay between integrated modelling and Bayesian analysis is discussed in Maunder (2003).

9.7.4 *Model Selection*

The general procedures discussed in Sect. 5.6 still apply but some specifics to the connected likelihood method are worth noting. Besbeas et al. (2002) combined ring-recovery data with census data. The number of age classes, and other model aspects, were those obtained from analysing the ring-recovery data alone. An alternative approach would be to base model selection on the connected likelihood. However for that kind of data combination, when selecting the number of age classes using AIC, Besbeas et al. (2014) found that it can be better to use recovery data alone.

9.7.5 *Integrated Modelling in Fisheries Research*

In this chapter, we have seen how state-space models and integrated population modelling allow the inclusion of information at the population and individual levels in a single framework. The state-space model for the population survey data opens the way to integrating different types of information and we have shown how data from multiple surveys can be simultaneously analysed in two different ways: by adding additional model structure in the model, as in Sect. 9.2, or by combining different likelihoods for different surveys with shared parameters for common processes, as in Sect. 9.3. The focus in the chapter has been on the latter and its application in population ecology, but integrated analysis has a long history in fisheries, dating back to Fournier and Archibald (1982), with developments in e.g. Maunder (1998, 2001, 2003). A recent review of integrated analysis in fisheries research is provided by Maunder and Punt (2013). Hoyle and Maunder (2004) provide a non-fisheries example, northeastern offshore spotted dolphin *Stenella attenuata*. Maunder (2004) uses IPM to carry out Population Viability Analysis (see Sect. 4.2).

One of the highly generalized integrated analysis models in fisheries is termed stock synthesis (Methot and Wetzel 2013), in which different data sets often contain contradictory information, and one of the main difficulties is determining the appropriate weighting factor between the different data types. In such cases, sensitivity analyses that investigate the influence of the weighting factors are an important part of the assessment. There are thus interesting analogies but also differences between integrated analysis in population ecology in general and particular applications in fisheries research and management. These differences stem from the different types of data, processes such as catchability, selectivity and aging imprecision, but also modelling purpose and complexity. However the underlying logic is the same, and research in both areas would benefit from better connection.

Chapter 10
Concluding Remarks

Modelling population dynamics is challenging because of multiple sources of uncertainty and variation. With rare exceptions, the number of individuals in a population cannot be completely censused. Instead one typically has incomplete counts, samples, or imperfect measurements of the population. These raw data may be mathematically summarized to provide estimates or indices of abundance, survival, reproduction, movement, growth. In either case, inferences about the population based on the raw data and the mathematical summaries will contain sampling error, i.e. uncertainty about the true state of the population (*partial observability*, Williams et al. 2001). However, even if the population were completely and perfectly known over a sequence of points in time, any mathematical model for the dynamics and subsequent predictions made by the model will be imperfect and contain *structural uncertainty*, *parameter uncertainty*, and *process variation* (Williams et al. 2001). Structural uncertainty is uncertainty about the underlying average or expected dynamics. For example the deterministic structure might be modelled by a simple Beverton–Holt model, but the real average is more complex. Even if the general form of the deterministic structure is correctly specified by a Beverton–Holt model, model parameters still need to be estimated, i.e. there will be parameter uncertainty. Finally even given the correct deterministic structure and perfectly known parameter values, there is inherent random or natural variation, such as environmental variation (e.g. between-year variation in reproductive rates) and demographic stochasticity (e.g. if expected reproductive rate was 2 progeny per female per year, there is between-female variation in the number of progeny, say 0, 1, 2, or 3 progeny).

State-space models provide a powerful, unifying framework for modelling population dynamics, giving a structure for explicitly accounting for each of the above sources of uncertainty. The observation model defines the relationship between the sample data and the true, but unknown state of the system and quantifies the degree of uncertainty in the relationship. Multiple state process models can be postulated and thereby one can make explicit the structural uncertainty, allowing for the likelihood that the truth is not in that set. The degree of structural uncertainty can be quantified by various measures of relative predictive performance or goodness

K.B. Newman et al., *Modelling Population Dynamics: Model Formulation, Fitting and Assessment using State-Space Methods*, Methods in Statistical Ecology, DOI 10.1007/978-1-4939-0977-3_10, of fit. Environmental stochasticity can be described via hierarchical state-space models, allowing vital rates, for example, to vary between years and/or to be modelled as a function of environmental conditions. Demographic stochasticity is characterized by the variation in the state process model. Parameter uncertainty can be quantified by standard errors or posterior distributions for parameter estimates.

Translating this general framework into one or more concrete SSMs for a specific case begins by clearly articulating the purpose of the model. Model purpose will guide the steps of formulation, fitting and assessment of SSMs. The model should be designed to fulfill that purpose, but should be no more complex than necessary (Starfield and Bleloch 1991). For example, if the purpose is to predict the effect of a change in hunting regulations for female red deer on population growth rates, one might model reproduction rates as a function of some measure of hunting as a covariate. The above three steps do not necessarily proceed in this sequence; e.g. after formulating a model, one might learn that it cannot be fitted to the available data and a new formulation is required. Nor is a single iteration necessarily adequate; for example, assessment might indicate that none of the models are adequate for the desired purpose and new models must be formulated. In all cases, model formulation requires, at a minimum, that decisions be made about (a) spatial-temporal scope, e.g. geographic region and years; (b) temporal resolution, e.g. how finely time is partitioned, annually, seasonally, or daily; (c) spatial resolution, e.g. the geographic region might be split into three sub-regions; (d) biological resolution, e.g. the population is subdivided into subpopulations by sex, age, sexual maturity. Model purpose and the available data will guide decisions about each of these.

A key aspect of the state process model portion of the SSM is specification of the processes that drive population dynamics. In many applications, a simple model may suffice. For example, net rate of change in population abundance might be modelled as a function of one or more covariates, such as year, rainfall, temperature, etc. In this case, separate models for survival and birth are not defined. However, if we wish to understand how survival and birth rates are influenced by for example environmental variables, we need to model those processes separately. We may also wish to include further processes, such as movement, growth and sex assignment.

Formulation of the observation model portion of the SSM may also be guided by model purpose and will certainly be constrained by the available data; e.g. if gender information was not collected, then observations must be related to quantities formed by merging female and male components of the state vector. The degree to which the original field data are to be aggregated (if at all) prior to fitting a SSM is another model formulation decision. See Dennis et al. (2010) and Knape et al. (2013) for discussion of the handling of replicate data in the observation model.

Matrix models are a convenient and compact way of characterizing both the state process and the observation portions of a SSM. Directly writing down the state process model as a generalized Leslie or Lefkovitch matrix becomes increasingly infeasible as the complexity of the population dynamics increases. In Chap. 2, we show how each separate process can be expressed in matrix form, and in a building block fashion, one can obtain the generalized Leslie or Lefkovitch matrix simply through matrix multiplication. This makes the task of defining and building the dynamic processes of a complex population model much simpler, and less error-prone.

Matrix models themselves, however, are often just convenient approximations to more realistic process-specific distributional models. For example, if density dependence is included in survival or birth rates, the linearity assumed in matrix models may be violated. Thus we use the matrix models primarily as a tool for helping to define the model, and to relate the model to classical matrix population models; we use the specified distributions for the individual processes for model-fitting, and the approximation involved in specifying matrix models is therefore of no consequence. In Chap. 3, we show how these more realistic and flexible population models may be translated into state-space models.

Methods for fitting SSMs to time series of data are provided in Chap. 4, including both classical frequentist methods and Bayesian approaches. The methods vary considerably in terms of the degree of difficulty and technical expertise required. Software for the fitting is rapidly maturing but the task remains more technically demanding than fitting linear or generalized linear models.

While the tools provided in Chaps. 2–4 allow great flexibility in building models of population dynamics, there is no value in developing a complex model when the available data are very limiting. In Chap. 5, we address in more detail how to formulate an appropriate model, and we show how to assess the adequacy of the model. Methods are given for checking for parameter redundancy in state-space models. We also address model selection, model averaging and model diagnostics.

In Chap. 6, we show how to model population dynamics when closed-population assessment methods are used at each of several time points, yielding a time series of abundance estimates. In Chap. 7, we consider methods when the primary interest is in modelling survival rates from mark-recapture data. Chapter 8 also considers mark-recapture data, but the emphasis is now on abundance estimation, with an embedded population dynamics model. This has the advantage for example of ensuring that modelled birth rates are biologically plausible. Integrated population modelling, for which there are multiple data sources, is addressed in Chap. 9.

Advances in both technology and statistical methodology are opening up many options for modelling the dynamics of wild animal populations. Tags can provide detailed information on movement, on different activity types, and even on animal condition, generating large datasets to complement data from mark-recapture studies, sightings surveys, etc. Hidden Markov models offer sophisticated methods of modelling such data (King 2014). Population processes can be modelled flexibly as a function of covariates and/or random effects. The possibility of incorporating genetic modelling into the population dynamics models is intriguing, potentially allowing exploration of how a population adapts to changes in the environment. We believe that models will become increasingly more reliable for predicting the dynamics of populations subject to change, whether it is climate change, changes in land use, changes in harvesting or control strategies, introduction of habitat management for conservation, controlled reintroductions, etc. A key aspect of reliability is that the modelling methods should allow quantification of uncertainty, and hence of risk. Methods should routinely incorporate demographic stochasticity, measurement error and model uncertainty, which is possible adopting the approaches described in this book.

References

Akaike, H. (1973). Information theory and an extension of the maximum likelihood principle. In B. N. Petrov & F. Csaki (Eds.), *Second International Symposium on Information Theory* (pp. 267–281). Budapest: Akademiai Kiado.

Alho, J. M. (1990). Logistic regression in capture-recapture models. *Biometrics, 50*, 494–500.

Alspach, D. L., & Sorenson, H. W. (1972). Nonlinear Bayesian estimation using Gaussian sum approximations. *IEEE Transactions on Automatic Control, 17*, 439–448.

Anderson-Sprecher, R., & Ledolter, J. (1991). State-space analysis of wildlife telemetry data. *Journal of the American Statistical Association, 86*, 596–602.

Andrieu, C., Doucet, A., & Holenstein, R. (2010). Particle Markov chain Monte Carlo methods (with discussion). *Journal Royal Statistical Society, Series B, 72*, 269–342.

Anscombe, F.J. (1973). Graphs in Statistical Analysis. *American Statistician*, 27(1): 17–21

Arlot, S., & Celisse, A. (2010). A survey of cross-validation procedures for model selection. *Statistics Surveys, 4*, 40–79.

Arnason, A. N. (1973). The estimation of population size, migration rates and survival in a stratified population. *Research in Population Ecology, 15*, 1–8.

Baillie, S. R., & Green, R. E. (1987). The importance of variation in recovery rates when estimating survival rates from ringing recoveries. *Acta Ornithologica, 23*, 41–60.

Barker, R. J. (1997). Joint modelling of live-recapture, tag-resight, and tag-recovery data. *Biometrics, 53*, 666–677.

Barry, S. C., Brooks, S. P., Catchpole, E. A., & Morgan, B. J. T. (2003). The analysis of ring-recovery data using random effects. *Biometrics, 59*, 54–65.

Besbeas, P., Borysiewicz, R. S., Morgan, B. J. T. (2009). Completing the ecological jigsaw. In D. Thomson, E. G. Cooch, & M. J. Conroy (Eds.), *Modeling demographic processes in marked populations*. Environmental and ecological statistics (Vol. 3, pp. 513–539). New York: Springer

Besbeas, P., & Freeman, S. N. (2006). Methods for joint inference from panel survey and demographic data. *Ecology, 87*, 1138–1145.

Besbeas, P., Freeman, S. N., & Morgan, B. J. T. (2005). The potential of integrated population modelling. *Australian and New Zealand Journal of Statistics, 47*, 35–48.

Besbeas, P., Freeman, S. N., Morgan, B. J. T., & Catchpole, E. A. (2002). Integrating mark-recapture-recovery and census data to estimate animal abundance and demographic parameters. *Biometrics, 58*, 540–547.

Besbeas, P., Lebreton, J.-D., & Morgan, B. J. T. (2003). The efficient integration of abundance and demographic data. *Applied Statistics, 52*, 95–102.

Besbeas, P., McCrea, R. S., & Morgan, B. J. T. (2014). Integrated population model selection in ecology (in preparation).

K.B. Newman et al., *Modelling Population Dynamics: Model Formulation, Fitting and Assessment using State-Space Methods*, Methods in Statistical Ecology, DOI 10.1007/978-1-4939-0977-3,

Besbeas, P., & Morgan, B. J. T. (2012a). A threshold model for heron productivity. *Journal of Agricultural, Biological, and Environmental Statistics, 17*, 128–141.

Besbeas, P., & Morgan, B. J. T. (2012b). Kalman filter initialization for integrated population modelling. *Applied Statistics, 61*, 151–162.

Besbeas, P., & Morgan, B. J. T. (2014). Goodness of fit of integrated population models using calibrated simulation. *Methods in Ecology and Evolution* (In Revision).

Bishop, J. R. B. (2009). *Embedding population dynamics in mark-recapture models*. (Ph.D. thesis, University of St Andrews).

Bonner, S. J., Morgan, B. J. T., & King, R. (2010). Continuous covariates in mark-recapture-recovery analysis: A comparison of methods. *Biometrics, 66*, 1256–65.

Borchers, D. L., Buckland, S. T., & Zucchini, W. (2002). *Estimating animal abundance: Closed populations*. London: Springer.

Brandon, J. R., & Wade, P. R. (2006). Assessment of the Bering-Chukchi-Beaufort seas stock of bowhead whales using Bayesian model averaging. *Journal of Cetacean Research and Management, 8*, 225–239.

Brooks, S. P., Catchpole, E. A., Morgan, B. J. T., & Barry, S. C. (2000). On the Bayesian analysis of ring-recovery data. *Biometrics, 56*, 951–956.

Brooks, S. P., Friel, N., & King, R. (2003). Classical model selection via simulated annealing. *Journal of the Royal Statistical Society B, 65*, 503–520.

Brooks, S. P., & Gelman, A. (1998). Alternative methods for monitoring convergence of iterative simulations. *Journal of Computational and Graphical Statistics, 7*, 434–455.

Brooks, S. P., Gelman, A., Jones, G. L., & Meng, X.-L. (Eds.). (2011). *Handbook of Markov chain Monte Carlo*. Boca Raton: Chapman & Hall/CRC.

Brooks, S. P., King, R., Morgan, B. J. T. (2004). A Bayesian approach to combining animal abundance and demographic data. *Animal Biodiversity and Conservation, 27*, 515–529.

Brown, D. L. (2010). *Climate modelling for animal survival*. (Ph.D. thesis, University of Kent).

Buckland, S. T. (1980). A modified analysis of the Jolly-Seber capture-recapture model. *Biometrics*, 36, 419–435.

Buckland, S. T., Anderson, D. R., Burnham, K. P., Laake, J. L., Borchers, D. L., & Thomas, L. (2001). *Introduction to distance sampling: Estimating abundance of biological populations*. Oxford: Oxford University Press.

Buckland, S. T., Burnham, K. P., & Augustin, N. H. (1997). Model selection: An integral part of inference. *Biometrics, 53*, 603–618.

Buckland, S. T., Newman, K. B., Fernández, C., Thomas, L., & Harwood, J. (2007). Embedding population dynamics models in inference. *Statistical Science, 22*, 44–58.

Buckland, S. T., Newman, K. B., Thomas, L., & Koesters, N. B. (2004). State-space models for the dynamics of wild animal populations. *Ecological Modelling, 171*, 157–175.

Buoro, M., Prévost, E., & Gimenez, O. (2010). Investigating evolutionary trade-offs in wild populations of Atlantic salmon *Salmo salar*: Incorporating detection probabilities and individual heterogeneity. *Evolution, 64*, 2629–2642.

Burnham, K. P. (1993). A theory for combined analysis of ring-recovery and recapture data. In J. D. Lebreton & P. M. North (Eds.), *The study of bird population dynamics using marked individuals* (pp. 199–213). Berlin: Birkhauser Verag.

Burnham, K. P., & Anderson, D. R. (2002). *Model selection and inference: A practical information-theoretic approach* (2nd ed.). New York: Springer.

Carlin, B. P., & Louis, T. A. (1996). *Bayes and empirical bayes methods for data analysis*. London: Chapman & Hall.

Carlin, B. P., & Louis, T. A. (2009). *Bayesian methods for data analysis* (3rd ed.). Boca Raton: Chapman & Hall/CRC.

Carlin, B. P., Polson, N. G., & Stoffer, D. S. (1992). A Monte Carlo approach to nonnormal and nonlinear state-space modeling. *Journal of the American Statistical Association, 87*, 493–500.

Casella, G. (1985). An introduction to empirical Bayes data analysis. *The American Statistician, 39*, 83–87.

Caswell, H. (2001). *Matrix population models: Construction, analysis, and interpretation* (2nd ed.). Sunderland: Sinauer Associates.

Catchpole, E. A., Fan, Y., Morgan, B. J. T., Coulson, T. N., & Clutton-Brock, T. H. (2004). Sexual dimorphism, survival and dispersal in red deer. *Journal of Agricultural, Biological, and Environmental Statistics, 9*, 1–26.

Catchpole, E. A., Freeman, S. N., & Morgan, B. J. T. (1996). Steps to parameter redundancy in age-dependent recovery models. *Journal of the Royal Statistical Society B, 58*, 763–774.

Catchpole. E. A., Freeman, S. N., Morgan, B. J. T., & Harris, M. P. (1998). Integrated recovery/recapture data analysis. *Biometrics, 54*(1), 33–46.

Catchpole, E. A., Kgosi, P. M., & Morgan, B. J. T. (2001). On the near-singularity of models for animal recovery data. *Biometrics, 57*, 720–726.

Catchpole, E. A., & Morgan, B. J. T. (1991). A note on Seber's model for ring-recovery data. *Biometrika, 78*, 917–919.

Catchpole, E. A., & Morgan, B. J. T. (1996). Model selection in ring-recovery models using score tests. *Biometrics, 52*, 664–672.

Catchpole, E. A., & Morgan, B. J. T. (1997). Detecting parameter redundancy. *Biometrika, 84*, 187–196.

Catchpole, E. A., & Morgan, B. J. T. (2001). Deficiency of parameter-redundant models. *Biometrika, 88*, 593–598.

Catchpole, E. A., Morgan, B. J. T., Coulson, T., Freeman, S. N., & Albon, S. D. (2000). Factors influencing Soay sheep survival. *Applied Statistics, 49*, 453–472.

Catchpole, E. A., Morgan, B. J. T., & Freeman, S. N. (1998). Estimation in parameter redundant models. *Biometrika, 85*, 462–468.

Catchpole, E. A., Morgan, B. J. T., Freeman, S. N., & Peach, W. J. (1999). Modelling the survival of British lapwings, *Vanellus vanellus* using ring-recovery data and weather covariates. *Bird Study, 46*(Suppl), S5–S13.

Catchpole, E. A., Morgan, B. J. T., & Tavecchia, G. (2008). A new method for analysing discrete life-history data with missing covariate values. *Journal of the Royal Statistical Society B, 70*, 445–460.

Celeux, G., Forbes, C. P., Robert, C. P., & Titterington, D. M. (2006). Deviance information criteria for missing data models. *Bayesian Analysis, 1*, 651–674.

Choquet, R., & Cole, D. J. (2012). A hybrid symbolic-numerical method for determining model structure. *Mathematical Biosciences, 236*, 117–125.

Clark, J. S. (2005). Why environmental scientists are becoming Bayesians. *Ecology Letters, 8*, 2–14.

Clarke, E. D., Spear, L. B., McCracken, M. L., Marques, F. F. C., Borchers, D. L., Buckland, S. T., & Ainley, D. G. (2003). Application of generalized additive models to estimate size of seabird populations and temporal trend from survey data collected at sea. *Journal of Applied Ecology, 40*, 278–292.

Colchero F., Medellin, R. A., Clark, J. S., Lee, R., & Katul, G. G. (2009). Predicting population survival under future climate change: Density dependence, drought and extraction in an insular bighorn sheep. *Journal of Animal Ecology, 78*, 666–673.

Cole, D. J. (2012). Determining parameter redundancy of multi-state mark-recapture models for sea birds. *Journal of Ornithology, 152*(Suppl 2), 305–315.

Cole, D. J., & McCrea, R. S. (2012). *Parameter redundancy in discrete state-space and integrated models*. (Technical Report UKC/SMSAS/12/012). University of Kent.

Cole, D. J., & Morgan, B. J. T. (2010a). A note on determining parameter redundancy in age-dependent tag return models for estimating fishing mortality, natural mortality and selectivity. *Journal of Agricultural, Biological, and Environmental Statistics, 15*, 431–434.

Cole, D. J., & Morgan, B. J. T. (2010b). Parameter redundancy with covariates. *Biometrika, 97*, 1002–1005.

Cole, D. J., Morgan, B. J. T., Catchpole, E. A., & Hubbard, B.A. (2012). Parameter redundancy in mark-recovery models. *Biometrical Journal, 54*, 507–523.

Cole, D. J., Morgan, B. J. T., & Titterington, D. M. (2010). Determining the parametric structure of non-linear models. *Mathematical Biosciences, 228*, 16–30.

Commandeur, J. J. F., Koopman, S. J., & Ooms, M. (2011). Statistical software for state space methods. *Journal of Statistical Software, 41*, 1–18.

Conn, P. B., Diefenbach, D. R., Laake, J. L., Ternent, M. A., & White, G. C. (2008). Bayesian analysis of wildlife age-at-harvest data. *Biometrics, 64*, 1170–1177.

Cormack, R. M. (1964). Estimates of survival from the sighting of marked animals. *Biometrika, 51*, 429–438.

Coull, B. A., & Agresti, A. (1999). The use of mixed logit models to reflect heterogeneity in capture-recapture studies. *Biometrics, 55*, 294–301.

Coulson, T. N., Catchpole, E. A., Albon, S. D., Morgan, B. J. T., Pemberton, J. M., Clutton-Brock, T. H., Crawley, M. J., & Grenfell, B. T. (2001). Age, sex, density, winter weather and population crashes in Soay sheep. *Science, 292*, 1528–1531.

Crosbie, S. F., & Manly, B. F. J. (1985). Parsimonious modelling of capture-mark-recapture studies. *Biometrics, 41*, 385–398.

Cunningham, C. L. (2002). *Improved management of north east Atlantic mackerel, using Bayesian modelling methodologies*. (Ph.D. thesis, Imperial College, University of London).

de Jong, P. (1988). A cross-validation filter for time series models. *Biometrika, 75*, 594–600.

de Valpine, P. (2002). Review of methods for fitting time-series models with process and observation error and likelihood calculations for nonlinear, non-Gaussian state-space models. *Bulletin of Marine Science, 70*, 455–471.

de Valpine, P. (2003). Better inferences from population-dynamics experiments using Monte Carlo state-space likelihood methods. *Ecology, 84*, 3064–3077.

de Valpine, P. (2004). Monte Carlo state-space likelihoods by weighted posterior kernel density estimation. *Journal of the American Statistical Association, 99*, 523–534.

de Valpine, P., & Hilborn, R. (2005). State-space likelihoods for nonlinear fisheries time-series. *Canadian Journal of Fisheries and Aquatic Science, 62*, 1937–1952.

Dennis B., Ponciano, J. M., Lele, S. R., Taper, M. L., & Staples, D.F. (2006). Estimating density dependence, process noise, and observation error. *Ecological Monographs, 76*, 323–341.

Dennis, B., Ponciano, J. M., & Taper, M. L. (2010). Replicated sampling increases efficiency in monitoring biological populations. *Ecology, 91*, 610–620.

Doucet, A., de Freitas, N., & Gordon, N. (Eds.). (2001). *Sequential Monte Carlo methods in practice*. New York: Springer.

Dupuis, J. A. (1995). Bayesian estimation of movement and survival probabilities from capture-recapture data. *Biometrika, 82*, 761–772.

Dupuis, J. A., Badia, J., Maublanc, M., & Bon, R. (2002). Survival and spatial fidelity of mouflons (*Ovis gmelini*): A Bayesian analysis of an age-dependent capture-recapture model. *Journal of Agricultural, Biological, and Environmental Statistics, 7*, 277–298.

Durbin, J., & Koopman, S. J. (2012). *Time series analysis by state space methods*, 2nd Edition. Oxford: Oxford University Press.

Eaton, M. A., Brown, A. F., Noble, D. G., Musgrove, A. J., Hearn, R., Aebischer, N. J., Gibbons, D. W., Evans, A., & Gregory, R. D. (2009). Birds of conservation concern 3: The population status of birds in the United Kingdom, Channel Islands and the Isle of Man. *British Birds, 102*, 296–341.

Edwards, D., & Havránek, T. (1985). A fast procedure for model search in multidimensional contingency tables. *Biometrika, 72*, 339–351.

Engle, R., & Watson, M. (1981). A one-factor multivariate time series model of metropolitan wage rates. *Journal of the American Statistical Association, 76*, 774–781.

Evans, N. D., & Chappell, M. J. (2000). Extensions to a procedure for generating locally identifiable reparameterisations of unidentifiable systems. *Mathematical Biosciences, 168*, 137–159.

Fan, Y., & Sisson, S. A. (2011). Reversible jump Markov chain Monte Carlo. In S. P. Brooks, A. Gelman, G. L. Jones, & X.-L. Meng (Eds.), *Handbook of Markov chain Monte Carlo* (pp. 69–94). Boca Raton: Chapman & Hall/CRC.

Fearnhead, P. (2011). MCMC for state space models. In S. P. Brooks, A. Gelman, G. L. Jones, & X.-L. Meng (Eds.), *Handbook of Markov chain Monte Carlo* (pp. 513–529) Boca Raton: Chapman & Hall/CRC.

Fewster, R. M., Buckland, S. T., Siriwardena, G. M., Baillie, S. R., & Wilson, J. D. (2000). Analysis of population trends for farmland birds using generalized additive models. *Ecology, 81*, 1970–1984.

Fournier, D., & Archibald, C. P. (1982). A general theory for analyzing catch at age data. *Canadian Journal of Fisheries and Aquatic Sciences, 39*, 1195–1207.

Freeman, S. N., & Besbeas, P. (2012). Quantifying changes in abundance without counting animals: extensions to a method of fitting integrated population models. *Journal of Ornithology, 152* (Suppl 2), S409–418.

Freeman, S. N., & Morgan, B. J. T. (1992). A modelling strategy for recovery data from birds ringed as nestlings. *Biometrics, 48*, 217–236.

Frühwirth-Schnatter, S. (1996). Recursive residuals and model diagnostics for normal and non-normal state space models. *Environmental and Ecological Statistics, 3*, 291–309.

Gelman, A., Carlin, J. B., Stern, H. S., & Rubin, D.B. (2003). *Bayesian data analysis*. Boca Raton: Chapman & Hall/CRC.

Gelman, A., & Meng, X. (1996). Model checking and model improvement. In W. R. Gilks, S. Richardson, & D. J. Spiegelhalter (Eds.), *Markov Chain Monte Carlo in Practice* (pp. 189–201). London: Chapman & Hall.

Geyer, C. (1996). Estimation and optimization of functions. In W. R. Gilks, S. Richardson, & D. J. Spiegelhalter (Eds.), *Markov chain Monte Carlo in practice* (pp. 241–258). London: Chapman & Hall.

Gilks, W. R., Richardson, S., & Spiegelhalter, D. J. (Eds.). (1996). *Markov chain Monte Carlo in practice*. London: Chapman & Hall.

Gimenez, O., Barbraud, C., Crainiceanu, C., Jenouvrier, S., & Morgan, B. J. T. (2006a). Semiparametric regression in capture-recapture modelling. *Biometrics, 62*, 691–698.

Gimenez, O., & Choquet, R. (2010). Individual heterogeneity in studies on marked animals using numerical integration: Capture-recapture mixed models. *Ecology, 91*, 148–154.

Gimenez, O., Covas, R., Brown, C. R., Anderson, M. D., Bomberger Brown, M., & Lenormand, T. (2006b). Nonparametric estimation of natural selection on a quantitative trait using mark-recapture data. *Evolution, 60*, 460–466.

Gimenez, O., Grégoire, A., & Lenormand, T. (2009a). Estimating and visualizing fitness surfaces using mark-recapture data. *Evolution, 63*, 3097–3105.

Gimenez, O., Morgan, B. J. T., & Brooks, S. P. (2009b). Weak identifiability in models for mark-recapture-recovery data. In D. L. Thomson, E. G. Cooch, & M. J. Conroy (Eds.), *Modelling demographic processes in marked populations*. Environmental and ecological statistics (Vol. 3, pp. 1057–1070). Springer, New York.

Gimenez, O., Rossi, V., Choquet, R., Dehais, C., Doris, B., Varella, H., Vila, J-P., & Pradel, R. (2007). State-space modelling of data on marked individuals. *Ecological Modelling, 206*, 431–438.

Gimenez, O., Viallefont, A., Choquet, R., Catchpole, E. A., & Morgan, B. J. T. (2004). Methods for investigating parameter redundancy. *Animal Biodiversity and Conservation, 27*, 561–572.

Green, P. J. (1995). Reversible jump Markov chain Monte Carlo computation and Bayesian model determination. *Biometrika, 82*, 711–732.

Grewal, M. S., & Andrews, A. P. (2010). Applications of Kalman filtering in aerospace 1960 to the present. *IEEE Control Systems Magazine, 30*, 69–78.

Grosbois, V., Harris, M. P., Anker-Nilssen, T., McCleery, R. H., Shaw, D. N., Morgan, B. J. T., & Gimenez, O. (2009). Survival at multi-population scales using mark-recapture data. *Ecology, 90*, 2922–2932.

Harrison, P. J., Buckland, S. T., Thomas, L., Harris, R., Pomeroy, P. P., & Harwood, J. (2006). Incorporating movement into models of grey seal population dynamics. *Journal of Animal Ecology, 75*, 634–645.

Harrison, P. J., Hanski, I., & Ovaskainen, O. (2011). Bayesian state-space modeling of metapopulation dynamics in the Glanville fritillary butterfly. *Ecological Monographs, 81*, 581–598.

Hart, J. D. (1994). Automated kernel smoothing of dependent data by using time series cross-validation. *Journal of the Royal Statistics Society B, 56*, 529–542.

Harvey, A. C. (1989). *Forecasting, structural time series models and the Kalman filter*. Cambridge: Cambridge University Press.

Harvey, A. C., & Pierse, R. G. (1984). Estimating missing observations in economic time series. *Journal of the American Statistical Association, 79*, 125–131.

Harville, D.A. (1977). Maximum likelihood approaches to variance component estimation and to related problems. *Journal of the American Statistical Association, 72*, 320–338.

Hastie, T., Tibshirani, R., & Friedman, J. (2009). *The elements of statistical learning: Data mining, inference, and prediction* (2nd ed.). New York: Springer.

Hastings, W. K. (1970). Monte Carlo sampling methods using Markov chains and their applications. *Biometrika, 57*, 97–109.

Higgins, K., Hastings, A., Sarvela, J. N., & Botsford, L. W. (1997). Stochastic dynamics and deterministic skeletons: Population behavior of Dungeness crab. *Science, 276*, 1431–1435.

Hilborn, R., & Mangel, M. (1997). *The ecological detective: Confronting models with data*. Princeton: Princeton University Press.

Hilborn, R., & Walters, C. J. (1992). *Quantitative fisheries stock assessment: Choice, dynamics and uncertainty*. New York: Chapman & Hall.

Hoyle, S. D., & Maunder, M. N. (2004). A Bayesian integrated population dynamics model to analyze data for protected species. EURING proceedings. *Animal Biodiversity and Conservation, 27*, 247–266.

Hubbard, B. A., Cole, D. J., & Morgan, B. J. T. (2014). Parameter redundancy in capture-recapture-recovery models. *Statistical Methodology, 17*, 17–29.

Huggins, R. M. (1989). On the statistical analysis of capture experiments. *Biometrika, 76*, 133–140.

Hunter, C. M., & Caswell, H. (2009). Rank and redundancy of multistate mark-recapture models for seabird populations with unobservable states. In D. L. Thomson, E. G. Cooch, & M. J. Conroy (Eds.), *Modeling demographic processes in marked populations*. Ecological and environmental statistics series (Vol. 3, pp. 797–826). Springer, New York.

Ionides, E. L., Breto, C., & King, A. A. (2006). Inferences for nonlinear dynamical systems. *Proceedings of the National Academy of Sciences, USA, 103*, 18438–18443.

Jiang, H., Pollock, K. H., Brownie, C., Hightower, J. E., Hoenig, J. M., & Hearn, W. S. (2007). Age-dependent tag return models for estimating fishing mortality, natural mortality and selectivity. *Journal of Agricultural, Biological, and Environmental Statistics, 12*,177–194.

Jolly, G. M. (1965). Explicit estimates from capture-recapture data with both death and immigration: Stochastic model. *Biometrika, 52*, 225–247.

Jolly, G. M. (1982). Mark-recapture models with parameters constant in time. *Biometrics, 38*, 301–321.

Jonsen, I. D., Myers, R. A., & James, M. C. (2006). Robust hierarchical state-space models reveal diel variation in travel rates of migrating leatherback turtles. *Journal of Animal Ecology, 75*, 1046–1057.

Julier, S. J., Uhlmann, J. K., & Durrant-Whyte, H. (1995). A new approach for filtering nonlinear systems. In *Proceedings of the American Control Conference* (pp. 1628–1632).

Kalman, R. E. (1960). A new approach to linear filtering and prediction problems. *Transactions of ASME Journal of Basic Engineering, 82*, 35–45.

Kalman, R. E., & Bucy, R. S. (1961). New results in linear filtering and prediction theory. *Journal of Basic Engineering, Transactions of ASME, Series D, 83*, 95–108.

Kanyamibwa, S., & Lebreton, J.-D. (1992). Variation des effectifs de Cigogne blanche et facteurs de milieu: Un modèle démographique. In J.-L. Mériaux, A. Schierer, C. Tombal, & J.-C. Tombal (Eds.), *Les Cigognes d'Europe* (pp. 259–264). Metz: Institut Européen d'Ecologie.

Kass, R. E., & Raftery, A. E. (1995). Bayes factors. *Journal of the American Statistical Association, 90*, 773–795.

Kendall, W. L., Nichols, J. D., & Hines, J. E. (1997). Estimating temporary emigration using capture-recapture data with Pollock's robust design. *Ecology, 78*, 563–578.
Kéry, M., & Schaub, M. (2012). *Bayesian population analysis using WinBUGS*. Amsterdam: Academic.
King, R. (2011). Statistical ecology. In S. P. Brooks, A. Gelman, G. L. Jones, & X.-L. Meng (Eds.), *Handbook of Markov chain Monte Carlo* (pp. 419–447). Boca Raton: Chapman & Hall/CRC.
King, R. (2012). A review of Bayesian state-space modelling of capture-recapture-recovery data. *Interface Focus, 2*, 190–204.
King, R. (2014). Statistical ecology. *Annual Review of Statistics and its Application, 1*, 401–426.
King, R., & Brooks, S. P. (2002). Bayesian model discrimination for multiple strata capture-recapture data. *Biometrika, 89*, 785–806.
King, R., & Brooks, S. P. (2003). Closed-form likelihoods for Arnason-Schwarz models. *Biometrika, 90*, 435–444.
King, R., & Brooks, S. P. (2004). A classical study of catch-effort models for Hector's dolphins. *Journal of the American Statistical Association, 99*, 325–333.
King, R., Brooks, S. P., Mazzetta, C., Freeman, S. N., & Morgan, B. J. T. (2008). Identifying and diagnosing population declines: A Bayesian assessment of Lapwings in the UK. *Applied Statistics, 57*, 609–632.
King, R., Brooks, S. P., Morgan, B. J. T., & Coulson, T. (2006). Factors influencing Soay sheep survival: A Bayesian analysis. *Biometrics, 62*, 211–220.
King, R., & McCrea, R. S. (2014). A generalised likelihood framework for partially observed capture-recapture-recovery models. *Statistical Methodology, 17*, 30–45.
King, R., Morgan, B. J. T., Gimenez, O., & Brooks, S. P. (2009). *Bayesian analysis for population ecology*. London: Chapman & Hall/CRC.
Kirkpatrick, S. (1984). Optimization by simulated annealing: Quantitative studies. *Journal of Statistical Physics, 34*, 975–986.
Kitagawa, G. (1987). Non-Gaussian state-space modeling of nonstationary time series. *Journal of the American Statistical Association, 82*, 1032–1041.
Knape, J., Besbeas, P., de Valpine, P. (2013). Using uncertainty estimates in analysis of population time series. *Ecology, 94*, 2097–2107.
Laake, J. L., Punt, A. E., Hobbs, R., Ferguson, M., Rugh, D., & Breiwick, J. (2013). Gray whale southbound migration surveys 1967-2007: An integrated re-analysis. *Journal of Cetacean Research and Management, 12*, 287–306.
Lahoz-Monfort, J. J., Morgan, B. J. T., Harris, M. P., Wanless, S., & Freeman, S. N. (2011). A capture-recapture model for exploring multi-species synchrony in survival. *Methods in Ecology and Evolution, 2*, 116–124.
Langrock, R., & King, R. (2013). Maximum likelihood estimation of mark-recapture-recovery models in the presence of continuous covariates. *Annals of Applied Statistics, 7*, 1709–1732.
Lebreton, J. D. (1973). Introduction aux modèles mathématiques de la dynamique des populations. *Informatique et Biosphère, Paris*, 77–116.
Lebreton, J-D., Burnham, K. P., Clobert, J., & Anderson, D. R. (1992). Modeling survival and testing biological hypotheses using marked animals: A unified approach with case studies. *Ecological Monographs, 62*, 67–118.
Lebreton, J. D., & Isenmann, P. (1976). Dynamique de la population Camarguaise de mouettes rieuses *Larus ridibundus* L.: Un modèle mathématique. *Terre et Vie, 30*, 529–549.
Lefkovitch, L. P. (1965). The study of population growth in organisms grouped by stages. *Biometrics, 21*, 1–18.
Leslie, P. H. (1945). On the use of matrices in certain population mathematics. *Biometrika, 33*, 183–212.
Leslie, P. H. (1948). Some further notes on the use of matrices in population mathematics. *Biometrika, 35*, 213–245.
Liang, D. F., & Christensen, G. S. (1978). Estimation for discrete nonlinear time-delayed systems and measurements with correlated and coloured noise processes. *International Journal of Control, 28*, 1–10.

Link, W. A., & Barker, R. J. (2005). Modeling association among demographic parameters in analysis of open-population capture-recapture data. *Biometrics, 61*, 46–54.

Link, W. A., & Barker, R. J. (2009). *Bayesian inference with ecological applications*. San Diego: Academic.

Liu, J. S. (2001). *Monte Carlo strategies in scientific computing*. New York: Springer.

Liu, J. S., & Chen, R. (1998). Sequential Monte Carlo methods for dynamic systems. *Journal of the American Statistical Association, 93*, 1032–1044.

Liu, J., & West, M. (2001). Combined parameter and state estimation in simulation-based filtering. In A. Doucet, N. de Freitas, & N. Gordon (Eds.), *Sequential Monte Carlo methods in practice* (pp. 197–223). New York: Springer.

Lunn, D., Jackson, C., Best, N., Thomas, A., & Spiegelhalter, D. (2013). *The BUGS book: A practical introduction to Bayesian analysis*. Boca Raton: CRC Press.

Lunn, D. J., Thomas, A., Best, N., & Spiegelhalter, D. (2000). WinBUGS: A Bayesian modelling framework: Concepts, structure, and extensibility. *Statistics and Computing, 10*, 325–337.

Marchant, J. H., Hudson, R., Carter, S. P., & Whittington, P. A. (1990). *Population trends in British breeding birds*. Tring: British Trust for Ornithology.

Maunder, M. N. (1998). *Integration of tagging and population dynamics models in fisheries stock assessment*. (Ph.D. thesis, University Washington, Seattle).

Maunder, M. N. (2001). Integrated tagging and catch-at-age analysis (ITCAAN). In G. H. Kruse, N. Bez, A. Booth, M. W. Dorn, S. Hills, R. N. Lipcius, D. Pelletier, C. Roy, S. J. Smith, & D. Witherell (Eds.), *Spatial processes and management of fish populations* (pp. 123–146, Alaska Sea Grant College Program Report No. AK-SG-01-02). University of Alaska Fairbanks.

Maunder, M. N. (2003). Paradigm shifts in fisheries stock assessment: From integrated analysis to Bayesian analysis and back again. *Natural Resource Modeling, 16*, 465–475.

Maunder, M. N. (2004). Population viability analysis based on combining Bayesian, integrated, and hierarchical analyses. *Acta Oecologica, 26*, 85–94.

Maunder, M.N., & Deriso, R.B. (2011). A state-space multistage life cycle model to evaluate population impacts in the presence of density dependence: illustrated with application to delta smelt (*Hypomesus transpacificus*). *Canadian Journal of Fisheries and Aquatic Sciences, 68*, 1285–1306.

Maunder, M. N., & Punt, A. E. (2013). A review of integrated analysis in fisheries stock assessment. *Fisheries Research, 142*(SI), 61–74.

McCarthy, M. (2007). *Bayesian methods for ecology*. Cambridge: Cambridge University Press.

McCrea, R. S., & Morgan, B. J. T. (2011). Multistate mark-recapture model selection using score tests. *Biometrics, 67*, 234–241.

McCrea, R. S., & Morgan, B. J. T. (2014) . *Analysis of capture-recapture data*. London: Chapman & Hall/CRC.

McCrea, R. S., Morgan, B. J. T., & Bregnballe, T. (2012a). Model comparison and assessment for multi-state capture-recapture-recovery models. *Journal of Ornithology, 152*(Suppl 2), S293–S303.

McCrea, R. S., Morgan, B. J. T., Brown, D. I., & Robinson, R. A. (2012b). Conditional modelling of ring-recovery data. *Methods in Ecology and Evolution, 3*, 823–831.

McCrea, R. S., Morgan, B. J. T., Gimenez, O., Besbeas, P., Lebreton, J.-D., & Bregnballe, T. (2010). Multi-site integrated population modelling. *Journal of Agricultural, Biological, and Environmental Statistics, 15*, 539–561.

McCrea, R. S., Morgan, B. J. T., & Pradel, R. (2014). Diagnostic goodness-of-fit tests for joint recapture and recovery models. *Journal of Agricultural, Biological and Environmental Statistics*, DOI:10:1007/s13253-014-1074-1.

McCullagh, P., & Nelder, J. A. (1989). *Generalized linear models* (2nd ed.). London: Chapman & Hall.

Meinhold, R. J., & Singpurwalla, N. D. (1983). Understanding the Kalman filter. *American Statistician, 37*, 123–127.

Mendelssohn, R. (1988). Some problems in estimating population sizes from catch-at-age data. *Fishery Bulletin, 86*, 617–630.

Methot, R. D., & Wetzel, C. R. (2013). Stock synthesis: A biological and statistical framework for fish stock assessment and fishery management. *Fisheries Research, 142*, 86–99.

Metropolis, N., Rosenbluth, A. W., Rosenbluth, M. N., Teller, A. H., & Teller, E. (1953). Equation of state calculations by fast computing machines. *Journal of Chemical Physics, 21*, 1087–1092.

Meyer, R., & Millar, R. B. (1999). BUGS in Bayesian stock assessments. *Canadian Journal of Fisheries and Aquatic Science, 56*, 1078–1086.

Millar, R. B. (2009). Comparison of hierarchical Bayesian models for overdispersed count data using DIC and Bayes' factors. *Biometrics, 65*, 962–969.

Millar, R. B., & Meyer, R. (2000a). Bayesian state-space modeling of age-structured data: Fitting a model is just the beginning. *Canadian Journal of Fisheries and Aquatic Science, 57*, 43–50.

Millar, R. B., & Meyer, R. (2000b). Non-linear state space modelling of fisheries biomass dynamics by using Metropolis-Hastings within-Gibbs sampling. *Applied Statistics, 49*, 327–342.

Morgan, B. J. T. (2009). *Applied stochastic modelling* (2nd ed.). Boca Raton: Chapman & Hall/CRC.

Morgan, B. J. T., Catchpole, E. A., & Coulson, T. (2004). Conditional methodology for individual case history data. *Applied Statistics, 53*, 123–131.

Morgan, B. J. T., & Freeman, S. N. (1989). A model with first-year variation for ring-recovery data. *Biometrics, 45*, 1087–1102.

Morgan, B. J. T., Palmer, K. J., & Ridout, M. S. (2007). Negative score test statistic. *American Statistician, 61*, 285–288.

Morris, C. N. (1983). Parametric empirical Bayes inference: Theory and applications. *Journal of the American Statistical Association, 78*, 47–65.

Morris, W. F., & Doak, D. F. (2002). *Quantitative conservation biology*. Sunderland: Sinauer Associates.

Neter, J., Kutner, M. H., Nachtsheim, C. J., & Wasserman, W. (1996). *Applied linear statistical models* (4th ed.). Boston: McGraw-Hill.

New, L. F., Buckland, S. T., Redpath, S., & Matthiopoulos, J. (2011). Hen harrier management: Insights from demographic models fitted to population data. *Journal of Applied Ecology, 48*, 1187–1194.

New, L. F., Matthiopoulos, J., Redpath, S., & Buckland, S. T. (2009). Fitting models of multiple hypotheses to partial population data: Investigating the causes of cycles in red grouse. *American Naturalist, 174*, 399–412.

Newman, K. B. (1998). State-space modelling of animal movement and mortality with application to salmon. *Biometrics, 54*, 1290–1314.

Newman, K. B., Buckland, S. T., Lindley, S. T., Thomas, L., & Fernández, C. (2006). Hidden process models for animal population dynamics. *Ecological Applications, 16*, 74–86.

Newman, K. B., Fernández, C., Thomas, L., & Buckland, S. T. (2009). Monte Carlo inference for state-space models of wild animal populations. *Biometrics, 65*, 572–583.

Newman, K. B., & Lindley, S. T. (2006). Accounting for demographic and environmental stochasticity, observation error and parameter uncertainty in fish population dynamics models. *North American Journal of Fisheries Management, 26*, 685–701.

North, P. M., & Morgan, B. J. T. (1979). Modelling heron survival using weather data. *Biometrics, 35*, 667–682.

Pitt, M. K., & Shephard, N. (1999). Filtering via simulation: Auxiliary particle filters. *Journal of the American Statistical Association, 94*, 590–599.

Plummer, M. (2003). JAGS: A program for analysis of Bayesian graphical models using Gibbs sampling. In *Proceedings of the 3rd International Workshop on Distributed Statistical Computing* (pp. 20–22). Vienna.

Pohjanpalo, H. (1978). System identifiability based on the power series of the solution. *Mathematical Biosciences, 41*, 21–33.

Pollock, K. H. (1982). A capture-recapture design robust to unequal probability of capture. *Journal of Wildlife Management, 46*, 752–757.

Pollock, K. H. (2002). The use of auxiliary variables in capture-recapture modelling: An overview. *Journal of Applied Statistics, 29*, 85–102.

Pradel, R. (2005). Multievent: An extension of multistate capture-recapture models to uncertain states. *Biometrics, 61*, 442–447.

Punt, A. E., & Wade, P. R. (2010). *Population status of the Eastern North Pacific stock of gray whales in 2009*. NOAA Technical memorandum NMFS-AFSC-207. Retrieved October 25, 2013, from http://www.afsc.noaa.gov/publications/AFSC-TM/NOAA-TM-AFSC-207.pdf

Quinn, T. J., II, & Deriso, R. B. (1999). *Quantitative fish dynamics*. Oxford: Oxford University Press.

Rao, C. R. (1948). Large sample tests of statistical hypotheses concerning parameters with application to problems of estimation. *Proceedings of the Cambridge Philosophical Society, 44*, 50–57.

Rifflart, R., Marchand, F., Rivot, E., & Baglinière, J. L. (2006). Scale reading validation for estimating age from tagged fish recapture in a brown trout (*Salmo trutta*) population. *Fisheries Research, 78*, 380–384.

Rivot, E., & Prévost, E. (2002). Hierarchical Bayesian analysis of capture-mark-recapture data. *Canadian Journal of Fisheries and Aquatic Sciences, 59*, 1768–1784.

Rivot, E., Prévost, E., & Parent, E. (2001). How robust are Bayesian posterior inferences based on a Ricker model with regards to measurement errors and prior assumptions about parameters? *Canadian Journal of Fisheries and Aquatic Sciences, 58*, 2284–2297.

Rivot, E., Prévost, E., Parent, E., & Baglinière, J. L. (2004). A Bayesian state-space modelling framework for fitting a salmon stage-structured population model to multiple time series of field data. *Ecological Modelling, 179*, 463–485.

Robert, M., Faraj, A., McAllister, M. K., & Rivot, E. (2010). Bayesian state-space modelling of the De Lury depletion model strengths and limitations of the method, and application to the Moroccan octopus fishery. *ICES Journal of Marine Sciences, 67*, 1272–1290.

Rotella, J. J., Link, W. A., Nichols, J. D., Hadley, G. L., Garrott, R. A., & Proffitt, K. M. (2009). An evaluation of density-dependent and density-independent influences on population growth rates in Weddell seals. *Ecology, 90*, 975–984.

Royle, J. A. (2008). Modeling individual effects in the Cormack-Jolly-Seber model: A state-space formulation. *Biometrics, 64*, 364–370.

Royle, J. A., Dawson, D. K., & Bates, S. (2004). Modelling abundance effects in distance sampling. *Ecology, 85*, 1591–1597.

Royle, J. A., & Dorazio, R. M. (2008). *Hierarchical modeling and inference in ecology: The analysis of data from populations, metapopulations, and communities*. San Diego: Academic.

Royle, J. A., & Kéry, M. (2007). A Bayesian state-space formulation of dynamic occupancy models. *Ecology, 88*, 1813–1823.

Schaub, M., & Abadi, F. (2011). Integrated population models: A novel analysis framework for deeper insights into population dynamics. *Journal of Ornithology, 152*(Suppl 1), S227–S237.

Schaub, M., Gimenez, O., Sierro, A., & Arlettaz, R. (2007). Use of integrated modeling to enhance estimates of population dynamics obtained from limited data. *Conservation Biology, 21*, 945–955.

Schnute, J. (1994). A general framework for developing sequential fisheries models. *Canadian Journal of Fisheries and Aquatic Sciences, 51*, 1676–1688.

Schofield, M. R., & Barker, R. J. (2008). A unified capture-recapture framework. *Journal of Agricultural, Biological, and Environmental Statistics, 13*, 458–477.

Schofield, M. R., Barker, R. J., & MacKenzie, D. I. (2009). Flexible hierarchical mark-recapture modeling for open populations using WinBUGS. *Environmental and Ecological Statistics, 16*, 369–387.

Schwarz, C. J., & Arnason, A. N. (1996). A general methodology for the analysis of capture-recapture experiments in open populations. *Biometrics, 52*, 860–873.

Schwarz, C. J., Schweigert, J. F., & Arnason, A. N. (1993). Estimating migration rates using tag-recovery data. *Biometrics, 49*, 177–193.

Schwarz, G. E. (1978). Estimating the dimension of a model. *Annals of Statistics, 6*, 461–464.

Seber, G. A. F. (1965). A note on the multiple recapture census. *Biometrika, 52*, 249–259.

Seber, G. A. F. (1971). Estimating age-specific survival rates from bird-band returns when the reporting rate is constant. *Biometrika, 58*, 491–497.

Seber, G. A. F. (1973). *The estimation of animal abundance and related parameters*. New York: Hafner.

Seber, G. A. F. (1982). *The estimation of animal abundance and related parameters* (2nd ed.). New York: Macmillan.

Servanty, S., Choquet, R., Baubet, E., Brandt, S., Gaillard J.-M., Schaub, M., Togo, C., Lebreton, J.-D., Buoro, M., & Gimenez, O. (2010). Assessing compensatory vs. additive mortality using marked animals: A Bayesian state-space modeling approach. *Ecology, 91*, 1916–1923.

Shumway, R. A., & Stoffer, D. S. (1982). An approach to time series smoothing and forecasting using the EM algorithm. *Journal of Time Series Analysis, 3*, 253–264.

Skaug, H. J. (2002). Automatic differentiation to facilitate maximum likelihood estimation in nonlinear random effects models. *Journal of Computational and Graphical Statistics, 11*, 458–470.

Skaug, H. J., & Fournier, D. (2006). Automatic approximation of the marginal likelihood in non-Gaussian hierarchical models. *Computational Statistics and Data Analysis, 51*, 699–709.

Skaug, H. J., & Fournier, D. (2011). *Random effects in AD model builder*. ADMB-RE User Guide Version 10.0 (2011-01-18). Retrieved December 31, 2012, from http://www.admb-project.org/documentation/manuals/admb-user-manuals

Spiegelhalter, D. J., Best, N. G., Carlin, B. P., & Van der Linde, A. (2002). Bayesian measures of complexity and fit (with discussion). *Journal of the Royal Statistical Society B, 64*, 583–616.

Starfield, A. M., & Bleloch, A. L. (1991). *Building models for conservation and wildlife management* (2nd ed.). Edina: The Burgess Press.

Sullivan, P. J. (1992). A Kalman filter approach to catch-at-length analysis. *Biometrics, 48*, 237–258.

Swain, D. P., Jonsen, I. D., Simon, J. E., & Myers, R. A. (2009). Assessing threats to species at risk using stage-structured state-space models: Mortality trends in skate populations. *Ecological Applications, 19*, 1347–1364.

Tavecchia, G., Besbeas, P., Coulson, T., Morgan, B. J. T., & Clutton-Brock, T. H. (2009). Estimating population size and hidden demographic parameters with state-space modelling. *American Naturalist, 173*, 722–733.

ter Braak, C. J. F., van Strien, A. J., Meyer, R., & Verstrael, T. J. (1994). Analysis of monitoring data with many missing values: Which method? In W. Hagemeijer & T. Verstrael (Eds.), *Bird numbers 1992. Distribution, monitoring and ecological aspects* (pp. 663–673). *Proceedings of the 12th International Conference of the International Bird Census Council and European Ornithological Atlas Committee*. Beek-Ubbergeon, Sovon, The Netherlands.

Thomas, L., Buckland, S. T., Newman, K. B., & Harwood, J. (2005). A unified framework for modelling wildlife population dynamics. *Australian and New Zealand Journal of Statistics, 47*, 19–34.

Thomas, L., Burnham, K. P., & Buckland, S. T. (2004). Temporal inferences from distance sampling surveys. In S. T. Buckland, D. R. Anderson, K. P. Burnham, J. L. Laake, D. L. Borchers, & L. Thomas (Eds.), *Advanced distance sampling* (pp. 71–107) Oxford: Oxford University Press.

Tibshirani, R. (1996). Regression shrinkage and selection via the Lasso. *Journal of the Royal Statistical Society B, 58*, 267–288.

Tierney, L., & Kadane, J. B. (1986). Accurate approximations for posterior moments and marginal densities. *Journal of the American Statistical Association, 81*, 82–86.

Trenkel, V. M., Elston, D. A., & Buckland, S. T. (2000). Fitting population dynamics models to count and cull data using sequential importance sampling. *Journal of the American Statistical Association, 95*, 363–374.

Wade, P. R. (2002). A Bayesian stock assessment of the eastern Pacific gray whale using abundance and harvest data from 1967–1996. *Journal of Cetacean Research and Management, 4*, 85–98.

Wan, E. A., & Nelson, A. T. (2001). Dual extended Kalman filter methods. In S. Haykin (Ed.), *Kalman filtering and neural networks* (pp. 123–173). New York: Wiley.

Wan, E. A., & van der Merwe, R. (2001). The unscented Kalman filter. In S. Haykin (Ed.), *Kalman filtering and neural networks* (pp. 221–280). New York: Wiley.

Ward E. J., Chirakkal, H., Gonzalez-Suarez, M., Aurioles-Gamboa, D., Holmes, E. E., & Gerber, L. (2010). Inferring spatial structure from time-series data: Using multivariate state-space models to detect metapopulation structure of California sea lions in the Gulf of California, Mexico. *Journal of Applied Ecology, 47*, 47–56.

West, M., & Harrison, J. (1997). *Bayesian forecasting and dynamic models* (2nd ed.). New York: Springer.

White, G. C., & Burnham, K. P. (1999). Program MARK: Survival estimation from populations of marked animals. *Bird Study, 46*(Suppl), 120–139.

Williams, B. K., Nichols, J. D., & Conroy, M. J. (2001). *Analysis and management of animal populations*. San Diego: Academic.

Williams, B. K., Szaro, R. C., & Shapiro, C. D. (2007). *Adaptive management: The U.S. department of the interior technical guide*. Washington, D.C.: Adaptive Management Working Group, U.S. Department of the Interior.

Wilson, A. M., Vickery, J. A., & Browne S. J. (2001). Numbers and distribution of northern lapwings *Vanellus vanellus* breeding in England and Wales in 1998. *Bird Study, 48*, 2–17.

Zheng, C., Ovaskainen, O., Saastamoinen, M., & Hanski, I. (2007). Age-dependent survival analyzed with Bayesian models of mark-recapture data. *Ecology, 88*, 1970–1976.

Index

K.B. Newman et al., *Modelling Population Dynamics: Model Formulation, Fitting and Assessment using State-Space Methods*, Methods in Statistical Ecology, DOI 10.1007/978-1-4939-0977-3,

Printed by Printforce, the Netherlands